Tulane Studies in Philosophy
VOLUME X

Studies in Whitehead's Philosophy

TULANE UNIVERSITY
NEW ORLEANS
1961

This volume may be purchased for
$2.00, plus postage, from
the Department of Philosophy, Tulane University,
New Orleans 18, La.

ISBN-13: 978-90-247-0284-8   e-ISBN-13: 978-94-011-8135-8
DOI: 10.1007/978-94-011-8135-8

Copyright 1961 by Martinus Nijhoff, The Hague, Netherlands
Softcover reprint of the hardcover 1st edition 1961

TABLE OF CONTENTS

The essays in the present volume are inspired by the attempt to celebrate the influence of the philosophy of Alfred North Whitehead in this year of the centennial anniversary of his birth.

STUDIES IN WHITEHEAD'S PHILOSOPHY

LIST OF ABBREVIATIONS

In the following essays, citations to the works of White-head will be made with the following abbreviations.

OT *The Organisation of Thought*, London and Philadelphia, 1917.

PNK *An Enquiry Concerning the Principles of Natural Knowledge*, Cambridge, 1919.

CN *The Concept of Nature*, Cambridge, 1920.

SMW *Science and the Modern World*, New York, 1925.

S *Symbolism, Its Meaning and Effect*, New York, 1927.

PR *Process and Reality*, New York, 1929.

FR *The Function of Reason*, Princeton, 1929.

AI *Adventures of Ideas*, New York, 1933.

MT *Modes of Thought*, New York, 1938.

E *Essays in Science and Philosophy*, New York, 1947.

Quotations from *Science and the Modern World*, from *Symbolism, Its Meaning and Effect*, from *Process and Reality*, from *Adventures of Ideas*, and from *Modes of Thought* are by permission of The Macmillan Company; from *The Function of Reason* are by permission of the Princeton University Press; and from *Essays in Science and Philosophy* are by the permission of Philosophical Library. From the writings of Charles Hartshorne, permission to quote has been granted by the following publishers: Longmans, Green and Co., "The Compound Individual" in *Philosophical Essays for Alfred North Whitehead*; The Free Press of Glencoe, *Reality as Social Process*; The University of Chicago Press, *The Philosophy and Psychology of Sensation* (copyright 1934), and *Philosophers Speak of God* (copyright 1953); Harper and Brothers Publishers, *Beyond Humanism*, and *Man's Vision of God*; Yale University Press, *The Divine Relativity*.

KANT AND WHITEHEAD, AND
THE PHILOSOPHY OF MATHEMATICS

EDWARD G. BALLARD

The problem concerning the relation between mathematics and the human situation has been disputed since the Renaissance. It will be advantageous to narrow the problem down. There are two related questions: In what sense is mathematics dependent upon or independent of the human being? In what sense it is dependent upon or independent of the world? Clearly these are not problems within mathematics, but rather are problems about mathematics, and they belong, I suppose, as much to philosophy as to any other discipline. Kant and Whitehead are opposed in their respective ways of treating them. Kant's position is more acceptable today than is often thought, but still it does not seem to allow for mathematics which is pure in the modern sense. Whitehead's doctrine allows for both pure and applied mathematics, but the coherence of his views may be questioned. I want to discuss certain opinions of each of these two philosophers and then to indicate a view of the dependence of mathematics to which these discussions seem naturally to lead.

It is reasonable to begin by placing the two philosophers just mentioned in opposition to a doctrine which recently has received much attention. I refer to the conventionalism derived from Poincaré's writings. Poincaré held that the axioms of geometry have their source not in an à priori and not in experimentation; rather, they are conventions freely established by the mathematician within the limitations imposed by logic.[1] This conventionalist view, frequently ex-

---

[1] "*Les axiomes géométriques ne sont donc ni des jugements synthétiques* a priori *ni des faits expérimentaux.*

"Ce sont des *conventions*; notre choix, parmi toutes les conventions possibles, est guidé par des faits expérimentaux; mais il reste *libre* et n'est limité que par la nécessité d'éviter toute contradiction." H. Poincaré, *La Science et l'Hypothèse*, (Paris, 1943), p. 66, italics in the original.

pressed in a more extreme form than that held by its origi-
nator, is widely shared today.

It is sometimes said, for example, that mathematics is a
game played with a typewriter; it is founded upon arbitrarily
selected statements and proceeds according to rules of operation which the mathematician lays down upon his own
authority. This arbitrariness, I take it, means that there is
no reason within mathematics why a given set of postulates
and operations is to be preferred to another. The choice
among logical possibilities is arbitrary; it is restricted by no
à priori. The only requirement is the logical one; once the
selection has been made, the mathematician must stick coherently to it. It follows that mathematics has no intrinsic
relation to the world or to human experience; however, by
clever assignment of experiential values to mathematical
variables, predictions of the future course of experience can,
in fact, be made. These predictions can be justified pragmatically, although in no other way. It is not easy to object
to this viewpoint when it is held by a mathematician who is
eager to get back to his work. It is rather less easy to justify
it when it is held by a philosopher who aspires to a rational
grasp of the whole human experience. It is uncompromisingly
rejected by both Kant and Whitehead.

I. MATHEMATICS IN KANT

Let us determine the sense in which mathematics is an
autonomous discipline within Kant's philosophy.

It will be desirable to start by setting aside, quite briefly,
certain interpretations of Kant which are not useful for this
enterprise. Not least aside from the point is the often repeated
remark that Kant's whole view assumed the uniqueness of
Euclidean geometry; hence, it is completely discredited by
the discovery of non-Euclidean geometry. Some of Kant's
defenders take criticisms of this sort quite seriously and try
to extricate Kant from the supposed difficulty. For example,
Bollert points out [2] that Riemannian space contains Euclidean

---

[1] *Einstein's Relativitätstheorie und ihre Stellung im System der Gesamterfahrung*
(Dresden, 1921).

space as a differential element. That is, Euclidean space functions as a limit to which Riemannian space approximates over small areas. Thus, Euclidean space might be said to define Riemannian space. This view is clearly a *tour de force*. At best it shows that Kant accidentally agreed with certain modern views. Another procedure of this type consists in pointing out that Euclidean geometry is the geometry on the horosphere, a figure in Lobatchevskian geometry obtained by stretching an elastic circle, so to speak, to infinity in opposite directions.[1] In addition, two-dimensional Riemannian geometry is valid upon the surface of an Euclidean sphere. Evidently there is a relation among these geometries. It has, in fact, been demonstrated that there are theorems common to all three. Perhaps then, by another *tour de force*, Kant's "real meaning" might be interpreted as referring to that which these geometries have in common. It is to be recalled, as a point in favor of this view, that Kant was not without geometrical sophistication; he made mention of geometries of many dimensions in *Die Metaphysischen Anfangsgrunde der Naturwissenschaft*. He does not, however, develop these suggestions nor does he work out their consequences for his philosophical view. On the whole, efforts to read into Kant's writings knowledge which he could not have possessed, or likely did not have, or in any event did not use, are mistaken; they do not assist us to share in his grasp of philosophy.

Cassirer's way of determining the essentials of Kant's view is to generalize his stated position and to argue that this generalized doctrine is as valid as respects mathematics and the sciences today as its more specific expression was in Kant's own time. Space and time as intuitions, therefore, are interpreted generally as the relational orders of coexistence and succession.[2] This general intuition of coexistence, Cassirer holds, is non-metrical. Space, thus, is separated from the specifically Euclidean metric. Its metrical character is the consequence of a (logically) posterior, conceptual interpre-

[1] Cf. *The Elements of Non-Euclidean Geometry*, D. M. Y. Sommerville, (New York, 1958) p. 165.
[2] *Substance and Function in Einstein's Theory of Relativity*, (New York, 1953) p. 417f.

tation. Further, there is no epistemological reason why the choice of metric should not be the outcome of physical experimentation.

Reichenbach attempts to turn the tables on this interpretation by arguing that it renders the metric axioms of geometry non-intuitive and hence non-synthetic.[1] It is difficult to follow Reichenbach's reasoning in this matter. Kant did not argue that geometrical propositions are synthetic because they are intuitive; in fact, he notes that some propositions used in mathematics are intuitive and analytic (CPR B–16f). Rather, he points out, typically mathematical propositions are in fact synthetic and necessarily applicable to experience. Then he argues that these propositions can be understood to be both synthetic and necessarily applicable to experience only if their synthesis is the consequence, not of an arbitrary decision or act on the part of the mathematician, but rather of the same intuitive and cognitive powers which form experience. That is, concepts and forms of experience can be understood as necessarily applicable to the objects discerned in experience only if the conditions of the possibility of experience are the same as the conditions of the possibility of the objects of experience (CPR A158–B197; A111). Cassirer recognizes, as Reichenbach does not, that the necessity thus imputed to experiential knowledge is not a logical necessity but rather a kind of natural necessity, a "scientific necessity." This is the necessity for organizing experience in definite ways if we are to succeed in having sciences of persistent, ordered, natural objects.

This notion of "scientific necessity," based upon the intrinsic relation which Kant established between mathematical relations and possible experience, is rejected by those philosophers who hold that all necessity is analytic. They misinterpret, as it appears to me, Kant's distinction between analytic and synthetic. The point can be rendered clear by considering the Kantian meaning of analytic. An analytic proposition is one whose predicate must be thought in the subject (CPR A7–B10) or is provable by the law of contradiction alone (CPR

---

[1] *Modern Philosophy of Science* (New York, 1959) p. 28f.

A151–B191). We could, on these grounds, prove that a triangle contains three angles, that it is a plane figure, that it is extended, etc. . . ., for all these notions are contained in its definition. The concept of a triangle would be self-contradictory if the analysis turned out otherwise. It is the occupation of the philosopher to clarify concepts by this sort of analysis (A716–B744). A mathematical *object*, however, is not exhausted by its definition. It is rather, that object to which a true or real definition corresponds (A242n). This mathematical object can be empirical or pure (A718). It is in either case a content or manifold which may be unified by the unifying activity of thought. The point is that this manifold is other than and contains more than the conceptual definition. Just what it contains, Kant held, can be envisaged only by constructions of mathematical forms (A713ff–B741ff). Mathematical constructions in the (pure) intuition are those to which empirical particularities of a given figure are irrelevant. But the empirical intuition must conform to pure intuition. Thus, we can learn about the necessary quantitative properties of empirical objects from constructions in pure intuition. For example, the geometer learns that the sum of the internal angles of a triangle is equal to 180° not by analyzing the concept or definition of triangle, for it is in fact not contradictory to conceive this sum to be either greater or less than 180°, but he learns what this sum must be objectively by suitable constructions in intuition (A717f–B745f). It was Kant's decision to limit the notion of significance in mathematics to constructions of this objective sort.

The objective and public character of these constructions arises from their synthetic character. The problem of synthesis concerns the connection of an otherwise empty concept to a content. It concerns, for example, the connection of the defined concept "triangle" to the figure containing 180° as the sum of its interior angles or to some empirical triangular object. The synthesis is non-arbitrary or necessary when the combination of concepts in a judgment conforms to, or corresponds to, the combination of elements and properties of the intuitional content (A105). It is evident that this necessity is not logical, it is intuitional or contentual. That is, intuition is not amorphous nor is it conceptually organized. Rather, it is organizable in certain definite ways, limited by the forms native to our intuition. The permissible types of organization correspond exactly to the categories of thought. The function of the transcendental deduction is to show that thought (imagination and concepts) contains just the same ordering principles and relations which are to be found exhibited in intuition. That is, the understanding can bring intuition to concepts.

It is evident that the mathematician brings intuitions to concepts when he expresses what is contained in a construction in mathematical notions, for the mathematical notions are concepts (A714–B742) and a construction is an exhibition of a concept in pure intuition (A713–B741). Thus the mathematician's knowledge is gained by reason from the construction of concepts in pure intuition. Further, all empirical intuitions must conform to pure intuition; thus mathematical statements are necessarily applicable to the empirical world. This necessary applicability of mathematics is the consequence of the dependence of the mathematical form encountered in experience upon the formative activity of intuition.

It seems to me difficult to deny that the things which Kant is interested in distinguishing are indeed different. The statement that a triangle is a closed plane figure containing three and only three angles, and the statement that the sum of its interior angles is 180° are clearly non-equivalent statements. Both differ from a statement concerning the sum of the angles of some particular triangle. Kant's opposition between analysis and synthesis, and between pure and empirical intuition, is an attempt to render these three distinctions clear.

Only in the light of a grasp of Kant's epistemological purpose can his doctrine be correctly evaluated. It is doubtful that he was interested in the logical foundations or logical character of a mathematical system, although he certainly recognized that the logical and non-contradictory character of mathematics is an indispensible property of the science. Other properties which, in his doctrine, are decisive are contained in the notion of construction. This notion does not affirm that non-Euclidean theorems, for which Kant could have had no illustrative construction, are in any sense wrong. They may be logically correct. But I think the point is that they can not be said to exist mathematically. Mathematics underived from intuition and, consequently, irrelevant to phenomena would be, in Kant's opinion, "mere preoccupation with mental chimeras" (A157–B196). On this point he takes issue with Leibniz. For the latter, a real definition is also a construction, but this "construction" is only a demonstration of logical possibility. For Kant, on the other hand, a real definition or explanation is "not the one which renders merely the concept clear but rather the one which renders its objective reality clear" (A241n). And "objective reality" refers to the public and ordered world of things in space and time (A492ff). Mathematical relations, then, are defined by selecting from the logically possible relations those which are characterized by a necessary reference to experience. Thus, mathematically significant or existent objects are constructible objects, and these also refer necessarily to the phenomenal world. Kant is quite clear about this; he writes, "they (mathematical concepts) would mean [bedeuten] nothing at all if we were not always able to display their meaning in appearances, (i.e. in empirical objects)" (A240–B299). This stand is merely a special application of his more general contention that meaning always involves reference to an object; "without the latter, it (a concept) has no sense, and is quite without content" (A239–B298). In sum, the quarrel between Kant and some of his mathematical critics concerns not so much the nature of mathematics as the criteria of meaning. Kant took his stand upon the conviction that a concept without an objective reference was a concept of nothing and, hence, was without significance.

I think it evident that Kant is not interested in finding a basis for mathematics in the modern sense of deducing it from logic. Whitehead's definition of mathematics as "the science concerned with the logical deduction of consequences from the general premises of all reasoning" (E277) would not so much be rejected by Kant as found to be unsuccessful in showing the connection between mathematics and the real world in which meaning is to be found. He might even accept this definition as delimiting a general area within which genuine mathematics is located. He, however, is concerned to discover the specific nature of mathematical knowledge by determining the criterion of mathematical significance. This he found in the notions of synthesis, construction, and objectivity. That Kant failed to anticipate future successes in synthesizing non-Euclidean geometries with intuition does not vitiate the criterion to which he was led. It is, in fact, the criterion to which the physicist or natural philosopher would naturally appeal. Kant's is a physicist's view of mathematics; Whitehead somewhere remarks that Kant would have been a great physicist had not other concerns absorbed his energies. Perhaps he was lacking in the aesthetic appreciation of formal structure about which Whitehead speaks so eloquently (MT83f). We may, then, summarize this point by remarking that mathematics is applied mathematics; he himself has said, "the very value of mathematics (that pride of human reason) rests upon the fact that it lends guidance to reason in respect to natural order and lawfulness" (A464–B492; cf. B147).

For Kant, then, mathematics is an independent science only within an area delimited by a double set of criteria. It is dependent upon satisfying certain logical standards. Likewise, more importantly and more essentially, it must satisfy criteria derived from possible experience. It follows from the second point that the mathematics which is genuine must be referable to actual or possible (imaginable) experience. A view of this kind is undeniably narrow, even if it is not so confused as some on Kant's critics have supposed. It appears arbitrary to exclude pure mathematics, in the modern sense, from the sphere of genuine mathematics.

II. MATHEMATICS AND NATURE IN WHITEHEAD

If mathematical concepts are not to be imposed arbitrarily upon experience nor denied authenticity unless abstracted from experience, then perhaps experience itself may be conceived to select in some way the mathematical relations from

out the general field of logical possibility. In effect, Whitehead elects to contrast pure with applied mathematics in this
manner. He doubts that there is any such independent topic
as applied mathematics; rather, applied mathematics is a
definitely selected portion of mathematics. He writes, "We
have, in fact, presented to our senses a definite set of transformations forming a congruence group, resulting in a set of
measure relations which are in no respect arbitrary . . . The
investigation of . . . this special congruence group is a perfectly
definite problem, to be decided by experiment" (E265; cf
CN129). Whence, though, comes this confidence in the ex-
istential relevance of some mathematics? How shall this
confidence be justified?

Whitehead's early interest in this matter led him to develop
the notion of event and overlapping of events in such a way
as to elucidate the operations and concepts of measurement.
He perceived, however, that the notions of event and overlapping were complex philosophical ideas in need of analysis
(PR352). He moved thence to an examination of the whole
philosophic tradition and to a reconstruction of fundamental
concepts. One aspect of his purpose is to account for the
route by which mathematical structure comes to be present
in knowledge in the same general way in which this structure
becomes ingredient in any actual entity. His method of doing
so seems to be a compromise between something resembling a
Kantian conviction, which seeks at the minimum to be assured
that the mathematical structure found in the world should
be present initially in the knower, and a kind of realism which
impels him to account for the mathematics in the mind as a
result of its being initially in the world.

Whitehead recognized that modern philosophy may be
said to begin with the rejection of the traditional, realistic
metaphysics and the acceptance of the subjectivist principle.
This principle accepts the subject experiencing as metaphysi-
cally irreducible (PR 43, 254). Descartes announced this
principle, but it was first used explicitly and consistently by
Kant. Whitehead believes, however, that Kant made a
radical error in its application; Kant attempted to show how
the objective world issues from the subject (PR 236). He is

able to perform this feat only by limiting experience to cognitive experience (PR 235) and by assuming that the subject is endowed with the forms of thought which operate necessarily to organize the data of experience. The first limitation renders his philosophy incapable of dealing adequately with more primitive types of experience; the second is an open violation of the "ontological principle." The latter principle affirms that apart from actual entities there is nothing. Hence, any propositions or beliefs intended to refer to anything real must be derivable from or abstractible from actual entities (PR 58, 101f). Violation of this principle leads to the fallacy of misplaced concreteness. Now Kant's forms and categories are, one might argue, unaccountably there. They are not derivable from more general properties of actuality, rather they are posited as generating the experienced actualities. Kant himself may even be interpreted as admitting their inexplicable giveness (cf. CPR A15–B29). In order to make a more radical beginning of philosophy, Whitehead would return to the actual occasion existent in the fullest sense rather than to some abstract and hypothetical forms of thought. To do so he employs the subjectivist principle in a manner just the reverse of Kant. He would show how the subject emerges from the object. Thus an act of experience is conceived as transforming objectivity into subjectivity (PR 236). I should like to examine this Kantianism-in-reverse just sufficiently to see how the mathematical form characteristic of one actual occasion of experience is said to be repeated in another; then it may become evident how, given certain mathematical properties encountered in experience, we may reasonably expect and recognize their recurrence.

We shall begin with a brief consideration of the doctrine of prehensions and its relation to the notion of possible mathematical form, and then attempt to determine how mathematical form is communicated from entity to entity. It may then be possible to determine how mathematical form becomes known.

Whitehead uses the term 'object' as correlative to 'subject.' Every object is discoverable by analysis of a subject (PR 89, 252). Objects are the data of a prehension; that is, they are the data of a subject's transaction with the world. Conversely that which enters into the experience of a subject is the given object or datum. Mathematical form, then, so far as it is descriptive of or constitutive of the real world, must characterize objects and be discoverable by the analysis of subjects.

Mathematical forms are a multiplicity of eternal objects of the objective species (PR 446). An eternal object is a type of definiteness which is non-actual and which is capable of, but independent of, realization in actual occasions (PR 70). The analysis of an eternal object discloses only other eternal objects (PR 34). Thus, eternal objects are given as potentially realizable, but in themselves they are indifferent to any temporal exemplifications. Whitehead cites the multiplication table as an example of such objects (MT 93). Being non-actual, eternal objects can not effect their own realization in the world. They are dependent for becoming the actual form of definiteness of an occasion upon something else; this something else is in part antecedent actuality (PR 101). The eternal objects having ingression in past occasions limit the ingression of eternal objects in occasions developing out of the former. Thus, the general potentiality offered by eternal objects is limited or "conditioned" by the occasions preceding in time (PR 34; MT 121). In *Science and the Modern World* this limited potentiality is called the spatio-temporal continuum. "Primarily the spatio-temporal continuum is a locus of relational possibility, selected from the more general realm of systematic relationship. This limited locus of relational possibility expresses one limitation of possibility inherent in the general system of the process of realization" (Chap. X). The real possibility, inherent in a cosmic epoch, is also termed the extensive continuum.[1] It is the most general relational system exemplifiable in any actual entities. This extensive continuum is not specified to the point of including specific metric relations; it includes

---

[1] Later he defines the extensive continuum as "a complex of entities united by the various allied relationships of whole to part, and of overlapping so as to possess common parts, and of contact, and of other relationships derived from these primary relationships" (PR 102, cf 123). It is undivided but divisible. One may think of it as the set of perspectives relevant to the actual world. Its causal transmission to future occasions is discussed in *Adventures of Ideas* p. 251.

only the most general notions of space and time. It resembles the Kantian forms of sensibility, only it is more general, and it is said to be derivative from rather than constitutive of the ordered world (PR 112).

Much the same doctrine is to be discerned in simplified form in *Modes of Thought*. There mathematics is said to be "concerned with certain forms of process issuing into forms which are components for further process" (MT 126). These forms may be considered either as relevant or as not relevant to the process of realization (MT 128); the former consideration appears to allude to the doctrine of the extensive continuum, the latter to abstract mathematics.

One may raise the question why this cosmic epoch has just the existent, necessary relationships prescribed by the extensive continuum. The ontological principle provides part of the answer, for according to this principle, the present character of the world must be causally dependent upon its past character. Thus prior occasions limit the possibility relevant to the present. Whitehead recognizes the difficulty of finding any further explanation for the relevance of real potentiality, indeed of any eternal objects, to the world. He is forced to fall back for an explanation of this relationship, upon a kind of theology, according to which eternal objects are related through "the mind of God" to the actual world (PR 73, 78). Perhaps this metaphysical explanation is another way of accepting the relevance of eternal objects to actual occasions as "inherent" (SMW loc cit). At any rate, to the two à priori factors already introduced, eternal objects and the antecedent universe, this theological doctrine adds a third.[1]

Let us, now, accept that there are actual entities which actualize and atomize the continuum and which, in their mutual relations, illustrate its forms of order. How is the communication of these forms of order to be conceived? The theory of prehensions is intended to answer this question by

[1] There may be a tenuous functional analogy between the Kantian à priori forms of intuitions, categories of thought, and transcendental unity of apperception and the Whiteheadian extensive continuum, eternal objects, and primordial nature of God.

means of a conception of the actual entity as genetically analyzable into a system of prehensions each of which consists in a subject in process of feeling, perceiving, or prehending the given world (its data) in a certain way (PR 31, 55, 65f). The data, then, come to be present in the subject in a determinate manner. The notion of absorbing or including data within the constitution of the actual entity is both a crucial and a difficult notion. In fact, "The philosophy of organism is mainly devoted to the task of making clear the notion of 'being present in another entity'" (PR 79f, cf. 88f). Upon the success of this task depends the possibility of clarifying the inheritance of mathematical form by one occasion from another (PR 438); upon it also depends the understanding of the way in which the knower becomes aware of mathematical form ordering the items and aspects of his experience. Whitehead desired to maintain two points: (1) the subject may feel or "absorb" (PR 261) another actual entity, as in the case of physical causation or perception; or (2) it may include or absorb an eternal object.

Now any entity, whether functioning as subject or as object, is illustrative of some eternal object in any respect in which it is definite. It may be that both the entity causing and the entity feeling an effect are illustrative of the same eternal object. The eternal object is conceived precisely in such a manner as to allow for such a possibility. When the subjective form of the effect is determined by the same eternal object as the cause, then that eternal object is said to be functioning relationally. There then exists a partial conformation between two actual entities (PR 78, 180, 362ff). In such an instance, the cause is said to be transformed into the effect in a sense; that is, the cause as datum is not merely represented in the effect; rather, it enters into the constitution of the effect, or is "re-enacted" in it (PR 362ff, 374f). Thus, the character of the past is reproduced and accumulated in the present, and those eternal objects which define the pervasive order of the world are continuously inherited by each developing occasion. Recurrence is re-enaction. This point, I think, is fundamental in Whitehead's doctrine concerning the relevance of mathematics to the world or to experience. All the mathe-

matics yet invented or to be invented may be said to be part
of the realm of general or abstract possibility. A selection
within this realm is made by reason of the antecedent occasions of the world having just the definite character which
they do have. The occasions constituting the contemporary
world grow out of the world immediately preceding. By
reason of causal relations they inherit the pervasive characteristics of order of the antecedent occasions and modify
their own properties, as they seek satisfaction, within these
limitations.

The question must now be raised: how do we know that
the form of the datum is "re-enacted" in the subject which
physically prehends the datum? Is Whitehead merely re-
peating the ancient assumption that cause and effect are
similar to each other? I must admit that much of his doctrine
concerning this question appears to me to be a reaffirmation
of just this ancient conviction expressed in a novel and
complex manner. It is difficult to see that he adduces any
new principle in favor of this belief or that he finds any new
evidence for it; although, it is perhaps possible to interpret
him as holding that there is in principle empirical evidence
for the absorption and objectification of data at least among
the higher organisms (PR 286f). Also, his generalization of the
notions of feeling and mentality place further difficulties in
the way of accepting a belief that the same mathematical
form embodied in the datum is prehended by the subject.

A thorough discussion of Whitehead's epistemology would
require a lengthy analysis of several Whiteheadian notions,
e.g. conceptual valuation, reversion, transmutation, hybrid
feelings, propositional feelings, consciousness, and his theory
of error. These doctrines maintain that conceptual feelings or
mental experiences originate along with physical feelings.
But since mental experience may also generate other mental
experiences, none can be uncritically referred back to the
original data. It is quite impossible to discuss all these
doctrines in this paper. It will suffice, for our purposes, to
note that a critical link, that between physical and conceptual
feelings, will have to be accounted for if we are to understand
either how the form of the datum is reproduced in a pre-

hending subject, or how we can come to know that such a
reproduction has occurred.

Conscious perception requires the entertainment of an
eternal object in abstraction from the actual entity which it
is said to qualify. Since the eternal object is independent of
its realization, there is nothing in its nature to prevent such
abstraction. Whitehead holds that a subject's conceptual
feelings are its response, not to the determinate character of
an eternal object, but rather to an eternal object in respect
to its power to determine (PR 366f). A response of this kind
refers to the eternal object as abstracted from any particular
occasion of experience and considers it as a potentiality for
qualifying experience. Conceptual feelings form the mental
pole of an actual occasion; they are concerned with what the
datum is for the subject in question. Consequently, conceptual
prehensions are primarily valuational in function (PR 360).
They simplify by selecting some significant quality from out
the wealth of physically prehended data. Consciousness, however, arises as an element in the subjective form only of
relatively few complex entities and is the consequence of
integration and comparison of physical and conceptual
feelings (PR 371).

The relevant aspect of Whitehead's theory of conscious
perception concerns its symbolic functioning. The theory is
dependent in part upon the distinction between the two
modes of perception: (a) causal efficacy, which is the integration of physical and conceptual feelings referred to by the
phrase, "the sense of the withness of the body"; and (b)
presentational immediacy, which is the perception of the
world through the senses (PR 91). Symbolic reference occurs
when one sort of experience or perception stands for the other.
This substitution of perception in one mode for that in another
is possible if there are some elements common to both. Now
the same eternal object is said to be common to both; likewise,
the same locus in space may be common to both (PR II Chap.
VIII). Spatial locus, however, is present in each in different
ways. In the mode of causal efficacy, the sense of presence in
the extended world is indirect, vague, and primarily emotional.
The percipient senses the causal presences of the world in

himself as center of a region. But in visual sense-perception
he is aware of the world as vividly spread out from this center.
Both modes are involved in cognition. We *want* to measure
and alter the world because of our causal involvement in it.
We *can* do so with some precision because we can coordinate
its spatial and temporal aspects presented to the senses by
means of the techniques of measurement (PR 256ff). Though
our sense of presence in the world is the consequence of perception via causal efficacy, observational data and scientific
measurements are made in the mode of presentational immediacy.

Abstraction, or the selective emphasis upon certain elements felt with the body, takes place in the process of conceptually evaluating (up or down) the patterns transmitted
by way of physical feelings. Such valuations are transmuted
into an average associated with the initial feelings; then,
when the perception is in the mode of presentational immediacy, this average feeling is projected as a pervasive character
upon a region of the contemporary world. The region is then
felt as possessing that general character. Sense perception,
thus, is an instance of the integration of physical with conceptual feeling. If the perception is true, then there is some
conformity between the projected pattern and the object
perceived.

These two modes of perception, indicating an involvement
in the world and a conceptual reflection of this involvement,
are designed to provide a basis for our coming to be aware of
mathematical (and other) patterns characterizing the world
and re-enacted in our experience (cf. PR 260, 489). It thus
provides us with a means for distinguishing between mathematics in general and the mathematics of experience. A
mathematical. form may be said to be "in nature" if from
some perceptual experience that form may be abstracted.

The detail of the geometry which characterizes our epoch
may be got at by way of an analysis of the actual occasions
constituting the world. As early as the *Principles of Natural
Knowledge* Whitehead had indicated his interest in developing
a doctrine of the extensive character of nature, not by assuming discrete points and instants, but rather by beginning

with interconnected regions (PNK 4). This schema, completing the doctrine of the extensive continuum, is corrected and developed in *Process and Reality*. In the later writing, it accepts as basic the extensive region and the relations of extensive connection.

Part of the real essence of an actual occasion is its region or standpoint, its "here" and "now"; hence, its region is not divisible or changeable (PR 93f, 107f, 124). However, an actual entity may be prehended as a datum by another actual occasion. As such, the datum is given as extended and divisible. Such an entity may be divided "coordinately" into subregions, yielding potential parts of the region or extensive quanta (PR 435). Of course, the actual entity itself is not divided; for its mental pole is incurably one. Neither is it in physical time (PR 434). Only its physical pole is in physical time and is divisible. Thus, the sum of parts resulting from coordinate division never yields the original actual entity. This is to say that the parts resulting from such a division are externally, never internally related (PR 438). Such external relations, however, are sufficient for mathematics, for within a set of regions or of the subsets of a region the method of forming abstractive sets renders possible the definition of the elements (point, line, etc.) of a geometry expressive of the extensive order of the world. The extensive connections among such parts is developed by Whitehead in terms of whole-part, overlapping, and contact (PR 103, quoted above, n. 6). The outcome is the four-dimensional continuum of our epoch.

In order to work out this doctrine Whitehead was obliged to make use of notions which sharpen further the clavage between the mental and physical poles. I have indicated certain of these differences: the two poles are not in time and space in the same sense; also they differ radically with respect to divisibility. It becomes more and more difficult to see that Whitehead has in any sense solved or avoided the problems inherent in any effort to discover an intelligible linkage between the physical and the mental. Whitehead holds that his doctrine avoids the Cartesian dualism (cf. PR 376); nevertheless, he seems rather to have generalized than to have avoided it, for every entity turns out to possess both a physical and a mental pole. So long as this dualism persists, it would seem to be impossible to distinguish in a non-arbitrary fashion between abstract possibility and the mathematical form inherent in experience or in nature. Although Whitehead's method of extensive abstraction, used for eliciting and defining the mathematical elements in nature which are necessary for measurement, is consistent with his system, it does not appear to be derived from his system. Other methods and conceptual schemas are possible. This fact serves to emphasize the difference between the mental and physical poles and suggests their relative independence.

In respect to the connection between these two poles, we learn from Whitehead only that a datum may be said to be re-enacted by the prehending subject both physically and mentally, and that the two poles are inseparable and equally important. Precisely how "conceptual registration" of physical feeling occurs, though, or how conceptual feelings become transmuted to feeling truly qualifying a region of the world, or how we know that all these operations occur, is not very clear. It is easy to conclude that Whitehead contented himself with giving these difficulties categoreal status much as certain rationalist philosophers allowed themselves to dispose of philosophical problems by attributing their solution to God.

Whitehead's intention, however, is of some importance. He desired to show that the more complex modes of interaction with the world, such as those characterizing cognition, apply further and develop out of the notions which justify the conviction that mathematical form is objective and is inherited by one occasion from another. The advantage of this position lies in the generality of its principles and the number of different kinds of problems which they are set to solve. Its real weakness lies, to my mind, in the highly metaphorical language and ideas which are used to elucidate the reasons for his convictions. That one entity should "absorb" another or be present in or be re-enacted in another is a notion difficult to communicate clearly. It suggests a kind of digestion of one entity by another, a suggestion scarcely to be taken literally. More specifically, it seems to be open to the charge of making an inexplicable transition between the physical and mental poles of reality, a transition which generalizes but does not solve the Cartesian dualism.

Whitehead criticizes the rationalistic philosophies of a Descartes or a Kant for misplacing abstractions. Whitehead himself, however, appears to be no less open to a criticism which the rationalists were accustomed to direct against their medieval and Aristotelian opponents. This is the accusation of anthropomorphism, the fallacy of misplacing human nature. So heavily charged are Whitehead's metaphorical terms with the sense of human involvement in life processes that it is difficult to see how he can avoid such an accusation. In any event, his persuasive metaphors concerning absorption and re-enactment are no less difficult to accept that the Kantian belief in forms of sensibility. Perhaps it is these heavily loaded metaphors which, referred to a nature conceived in terms of physics, leave us with the impression of an unresolved dualism. At any rate, whether the better philosophical explanation of the presence of mathematical form in nature is to be got by using the subjectivist principle in the Kantian rather than in the Whiteheadian manner is still an unresolved issue. Whitehead, at least, is more modern than Kant in seeking to allow for more mathematics than is embodied in experience. Mathematical form characterizing an object is said to be re-enacted in a subject both at the physical and mental levels in such a manner as to provide a knowing subject with the basis for making true mathematical statements about the objects of experience. But in addition, mathematical statements may be elaborated far beyond that which is exemplified in nature. Such abstract patterns provide the data for aesthetic experience of a high order. Pure mathematics as well as applied mathematics is, thus, possible.

III. THE DEPENDENCE OF MATHEMATICS

We have noted that for certain modern schools of philosophy mathematics is independent of experience. Of course, for the conventionalist, as well as for both Kant and Whitehead, mathematics must satisfy logical criteria. Beyond this, how-

ever, experience, the human mind, or nature, is thought to
lay down no a priori restrictions upon the mathematics which
is to be considered to be genuine. Consequently, to the conventionalist, the findings of mathematicians are given an
application not because they are "found in nature" by the
investigator, but because they are imposed upon experience
to suit the convenience of the investigator. That is, observations are translated into mathematical expressions according to a dictionary, correlating numbers with observation,
a correlation which is in every respect arbitrary, except that
it is pragmatically justified. At the opposite pole Kant
maintains mathematical relations are constitutive of the
phenomenal world, indeed are meaningless unless thus constitutive, and are found there by the investigator. The pure
mathematics, which is not to be encountered in nature, is a
mere amusement, not genuine mathematics. Whitehead's
view presents an intermediate position. He rejects both the
conventional position holding that mathematics of itself is
equally applicable or inapplicable as well as the Kantian
conviction which holds that only applicable mathematics is
genuine. For reasons to be discovered not in reason but in
experience, some mathematics is applicable in this cosmic
epoch. Mathematics itself, having an eternal status, is independent of, yet relevant to, illustration in nature.

Now we have concluded that one of these two latter views
is narrow and the other is, in its detail, not coherent. Kant
begins with experience and finds upon analysis that the necessity in nature is the necessity of the human mind. Thus, he
finds that he must suppose the human mind, its forms and
categories, to be given; also he requires a sensible given. But
then the Kantian is disappointed to find that he is able to
account only for that mathematics which is actually descriptive of the world as he knows it. Thus the accidents of the
Kantian's history and times tend to circumscribe mathematics.

Whitehead might appear to achieve a greater degree of
impersonality in his view of mathematics. By utilizing certain
properties of the world, he achieves a greater generality than
Kant, as is illustrated by his inclusion of pure as well as applied

mathematics in his system. It is to be doubted, however, that he *achieves* this generality; rather, he begins with it. The question to my mind is whether or not he can move in a logical and coherent way from the generality of his categoreal scheme back again to the limitations and the unity of human existence and experience. Whitehead assumes the reality of eternal objects. Also he must presuppose the indefinite past of the cosmos; this his ontological principle demands. In addition he has to assume the existence and functioning of God in order to account for the origination and specific definiteness of actual occasions. There is in addition, the question, as I have tried to indicate, whether the metaphors of absorption and re-enaction ever bring him, philosophically speaking, from the physical and divisible aspects of the cosmos to the mental pole of reality, in particular to the human being who seeks to understand his having knowledge of the cosmos.

I can entertain no objection to the use of metaphors. Indeed, thinking in such terms is unavoidable in the process of discovery. The process of discovery is the hunt for premises, and Whitehead describes metaphysics precisely as the hunt for premises. But the metaphors which assist us to a more exact doctrine are not that doctrine. Whitehead's philosophy appears to me to remain excessively close to the anthromorphic beginnings of all thought.

Must we now, conclude that the philosophies of both Kant and Whitehead tend toward psychologism or anthropomorphism in their efforts to understand the nature of mathematics? Are both unable to provide an account of the dependent status of mathematics without some reference to the biography and personal preferences or cultural biases of the mathematician? Such psychologism is commonly held to be subjectivistic in a pejorative sense, except by certain of the conventionalists' views, which welcome it. Now there are indeed some aspects of mathematics which clearly indicate the personal touch of the mathematician.

The relevant point emerges if one inquires why mathematicians develop the kind of mathematics which they do develop. Consider that an imaginable operator such as $1/0$ is quite universally rejected by mathematicians. Can use of this operator be excluded on mathematical grounds alone? Indeed, it is curious that zero should be the one number in a multiplicative system which has no inverse. True, it would work havoc in such a system to allow operations by this "number." By its use one might show that any number is the equivalent of any other; thus the system would become trivial. But 'havoc' is not a mathematical concept; neither is 'triviality' in the sense in which it suggests insignificance or obviousness. If we are to adhere exclusively to the definition of mathematics as the science which draws necessary conclusions (a definition which Whitehead cites with approval), then the "number" 1/0 would appear to be acceptable.[1] This logical possibility, however, is universally rejected. The fact seems to be evident that the mathematicians' disapproval of a system containing the operator, 1/0, follows from some non-mathematical premise.

It is sometimes said that a standard of the acceptability of a mathematical system is its possession of certain properties, for example, the algebraic properties of closure, association, commutation, and the presence of identity and inverse operators. In order to retain such properties, certain operations must be excluded, for instance division by zero. Requirements such as these would seem, at first glance, to provide a genuinely and exclusively mathematical standard by which a mathematical system could be recognized. However, the mathematician will immediately observe that all of these requirements are not always demanded; one may point, for example, to non-commutative systems. It is natural to ask why any of these properties are demanded. A further answer to such a question seems to be impossible except in terms of the interests of mathematicians.

The same point may be made by a consideration of the notion of the theorem in mathematics. At the minimum, a theorem is a proved or provable statement. Mathematical proof can be defined impersonally and with some logical rigor.

---

[1] It is interesting that Leibniz interpreted the expression 0/0 to equal 1. Leibniz, *Philosophical Papers and Letters*, ed. and tr. by L. E. Loemker, (Chicago, 1956), vol. ii, p. 886f.

If a primitive proposition is transformed by application of one of the accepted rules of transformation, the new proposition is proved. Similarly, if a proved proposition is transformed by one of the accepted rules of transformation, the new proposition is proved, etc. However, not all proved or provable statements are in fact regarded as theorems. Here again, the determinative factor seems to be something non-logical. As a mathematician once remarked, "Mathematics is a fine art; the woods are full of theorems, but we prove only the pretty ones." Again we note, the mathematician's interest, however it be phrased, must be added to the logical notion of provability before we can say what a mathematical theorem is.

It is a matter of moment further to identify this interest. It will certainly be observed that the interest at least of practical mathematicians is formed by their cultural and ideological environment. This is a round-about way of noting that mathematics has come to be what it is owing to extra-mathematical demands made upon the mathematician and upon the latter's desire to satisfy these demands. It is not quite easy to determine the character of these extra-mathematical criteria of mathematics. But the notion of mathematical relevance as it functions in the engineer's practice suggests a way of approach to the matter.

How does the engineer, in fact, select the fitting type of mathematical expression for solving a practical problem? The probability of his hitting upon the appropriate formula by arbitrary choice is very small indeed. It is only reasonable to account for his consistent success by his peculiar talent for grasping the real possibility offered by the situation. No engineer would think of confining himself to a one dimensional geometry in calculating the trajectory of a rocket; rather, he considers what is really possible under the circumstances. He observes that the conditions of the problem require the assumption that certain pervasive features of the location and circumstances are observable and will persist; consequently, he makes his calculations in a three or four dimensional space. Two things are suggested by this illustration. One is that the ballistics expert very quickly takes the general situation into consideration, applies criteria, and out of the indefinitely large amount of mathematical notions available he selects the type which will be useful. That is, his prior grasp upon the exigencies of the situation and the possibilities really relevant to it placed limitations upon his search for applicable formulae. The second point worthy of note is that the pure mathematician makes use of criteria not radically dissimilar from the practical one's. The pure mathematician working with Caley numbers may share few of the enthusiasms of the ballistics expert, but both would agree that operation with $1/0$ yields a trivial system which, therefore, is to be rejected. The trivial system, however "pure" or even logical, is uninteresting. But again, triviality is a non-mathematical criterion of value. The reason for the agreement between the two kinds of mathematicians lies not in mathematics but rather in the interest which both take in real possibility. Significance, even in pure mathematics, is linked to real possibility. The difference between the pure and the practical mathematician is probably that the pure mathematician is more sensitive to the aesthetic experience which mathematics can provide than is the practical one. But aesthetic experience is experience.

Generally speaking, then, the question whether a mathematical system is genuinely mathematical is decided by reference to certain formal requirements and in addition by a more or less covert appeal to interests or to relevance to human experience. The pure differs from the practical mathematician in respect to the kind of experience which he values. But the evaluation of mathematics by appeal to aesthetic values is, no less than an appeal to practical values, an appeal beyond mathematics to certain aspects of the natural and human world. Both Whitehead and Kant in their different ways recognize this dependence of mathematics upon certain prior demands imposed by human experience. For both, the significance of mathematics is its human significance. For both, there is something non-logical yet prior to mathematics upon which its complete definition depends.

In this last section I have termed that extra-mathematical something upon which mathematics is dependent the mathematician's interest. The latter, in turn, was equated to relevance to human experience. It is the mathematician's interest which makes a non-logical choice among logical possibilities. This manner of speaking is innocuous and seems to suggest nothing more than a kind of cultural relativism which today is not unfamiliar. Neither is it very enlightening. To note that the standard which forms and directs the mathematician's interest is a combination of demands for developing the kind of formulae which are useful in scientific inquiry and for satisfying the mathematicians' aesthetic inclinations, is merely to record a fact. Such a fact would seem to be susceptible of further explanation, preferably one which would throw further light upon the nature of mathematics. Sociological analyses and descriptions of this fact are sociologically enlightening, but they may scarcely be expected to add any justification or, or explanation for, the standard which is said to distinguish genuine from unacceptable mathematics. To observe that mathematical interest is acquired in consequence of discipline in the mathematical institutions which preserve and communicate a sensitiveness to appropriate standards is merely to pass the buck. It is difficult in any case to say that the institution forms the mathematician rather than that the mathematician, in virtue of his attachment to certain standards, forms the institution. The suggestion which has been repeatedly made in the classical tradition in philosophy is that culture or institutions and the standards of which institutions are the guardians are both expressions of something else.

But of what else? It is at least evident that the culture is not master of the situation; it cannot, any more than the individual, act arbitrarily. The state legislators who are said to have legalized the equation of π to 3, rather than to the rather awkward figure, 3.1416. . ., did not quite make mathematical history. They did not transcend the transcendental. Their extra-mathematical criteria stood in judgment rather over them than over mathematics. Nevertheless, their contribution was probably not different in kind from the mathematician's legitimate and disciplined interest which, as I have sought to argue, is so notable a factor in forming the science as we know it. Perhaps one may assume, as C. S. Peirce did in a related matter, a sort of natural talent on the part of certain individuals to take the "right" mathematical interest. If so, then the existence and nature of this peculiar talent become an important object of inquiry. Kant and Whitehead directed their search in just this direction.

The conventionalist, too, may say that the mathematician must have a talent for making the profitable, arbitrary assumptions. But then the notion of arbitrariness, under this limitation, becomes quite difficult. Though there may be no logical reason for any given selection among alternative logical possibilities, there is need for making some such selection if there is to be any mathematics. For not all imaginable operations can be allowed, not all mathematical systems need have all mathematical properties; not all provable statements are accepted as theorems. The conventionalist is saying, as we interpret his meaning, that the requisite selection is to be made by the mathematician. This is to say, that a mathematical system does not only have to satisfy a logical a priori, but it must satisfy a personal one as well. We are led to conclude that the mathematician himself is regarded as prior to mathematics.

It is not usual, of course, to express the conventionalist position in this way. But I can not see what else is its outcome. The mathematician is the mathematical a priori. The advantage of expressing conventionalism in this manner is that an evaluation is intimated. For an evident property of this outcome is its circularity. The mathematician, we say, is prior to mathematics. But the mathematician is one who is trained in mathematics, and mathematics must be a definite and recognizable discipline if he is to be trained in it. Thus, if there is a mathematician, there may be mathematics; and if there is mathematics, there may be mathematicians. This we already knew.

We are returned to Kant and Whitehead for they do not reject this circle; rather, their philosophies seem to seek to render it wider. Neither accepts the unanalyzed mathematician as the mathematical a priori. Kant seeks in the mathematician some aspect of his thought or experience which is the *sine qua non* of mathematics as of his whole culture. Whitehead seeks for the same in certain properties of that which is experienced. Thus both avoid the amorphous and uncritical assumption of the whole mathematician as the prior element in his science. And thus each advances in the philosophically profitable direction; although, perhaps, "with painful steps and slow."

WHITEHEAD ON SYMBOLIC REFERENCE

ALAN B. BRINKLEY

This paper, which is an exposition of some of the chief points of Whitehead's theory of symbolic reference with occasional critical remarks and questions, was suggested by a remark in Professor Victor Lowe's fine monograph, "The Development of Whitehead's Philosophy", which appeared in *The Philosophy of Alfred North Whitehead*, edited by Paul Arthur Schilpp (Second edition, Tudor, New York: 1951). Professor Lowe writes: "The theory of symbolic reference has, if I am not mistaken, a very great importance, entirely apart from the role it plays in Whitehead's speculative construction" (Schilpp 102).

There is some question as to how much one should include in a study of the theory. It has usually been treated as a part of Whitehead's theory of perception, and as such it has a history which spans Whitehead's long philosophical development. Such a study is certainly a desideratum, but it is a subject for another book during the current Whitehead revival. Again, one might simply write a commentary to *Symbolism: Its Meaning and Effect*. To some extent, that is what in brevity has been done here. The development of the theory as set forth there has been taken as the basic text, and references to the earlier and later books have been made where they seemed illuminating.

It is essential to the purpose of the present paper to treat Whitehead's theory of perception, but this is not treated for its own sake; a much fuller treatment would be required for that. There have been few philosophers who have left behind a body of work in which it is more difficult to disengage a few ideas for discussion without 'murdering to dissect.' Whitehead's philosophy is organismic in more than one sense and at least some of the defects of the present paper result from treating Whitehead's ideas as if they could be simply located·

"The human mind is functioning symbolically when some components of its experience elicit consciousness, beliefs, emotions, and usages, respecting other components of its experience" (S 7f). This is what Whitehead offers as a formal definition of symbolism and he goes on to explain that by 'symbols' he understands the first set of components, and by 'meaning,' the second. The process of transference between the symbols and the meaning, he calls 'symbolic reference.' Although he calls symbolic reference 'blind' (PR 273) and maintains that the common ground between the meaning on which the symbolic reference depends can be stated apart from any recourse to the percipient, the symbolic reference itself "is the active synthetic element contributed by the nature of the percipient" (S 8). It is Whitehead's view that "every actual thing is synthetic: and symbolic reference is one primitive form of synthetic activity whereby what is actual arises from its given phases" (S 20). Furthermore, the actual world, as it is for us, is the result of symbolic reference (S 18).

Experience untinted by symbolic reference is entirely an analytic ideal and all actual experience is colored by symbolic reference: "Complete ideal purity of perceptive experience, devoid of any symbolic reference, is in practice unobtainable . . ." (S 54). Another way of emphasizing this is to point out that "symbolic reference is the interpretive element in human experience" (PR 263) and that "our habits, our states of mind, our modes of behavior, all presuppose this 'interpretation' "(AI 279). Whitehead further develops the concept of 'interpretation' by saying that it always depends upon some common factor which grounds the transition from one mode of experience to another, as, for example, the transition from a tribal dance (as symbol) to a fertility myth (as meaning). The common factor can be said to be the reason for the transition from one mode of experience to another and if this transition is reversible, then "each pattern interprets the other as expressive of that common factor" (AI 321).

We can see, then, that whenever there is an occasion of

human experience, we can expect to find at least two sets of components with a highly variable, but nevertheless objective relation between them. It is the nature of the percipient which realizes the symbolic reference between the symbol set of components and the meaning set. It is the nature of the percipient and the unique character of that particular act of experience which determine for the occasion of experience those components which act as symbol and those which act as meaning. It is a fundamental view of Whitehead that no components of experience are inherently symbols or inherently meanings. Thus, considered in themselves, the components in no way require that there be a symbolic reference at all, nor do they require, if there is a symbolic reference, which direction the reference shall take.

While Whitehead insists that there must be some common ground, or structural element, shared by the two modes of experience if there is to be a symbolic reference at all, the common elements are not determinative for the reference. The two modes of experience or "schemes of presentation have structural elements in common, which identify them as schemes of presentation of the same world" but there are "gaps in the determination of the correspondence between the two morphologies. The schemes only partially intersect, and their true fusion is left indeterminate. The symbolic reference leads to a transference of emotion, purpose, and belief, which cannot be justified by an intellectual comparison of the direct information derived from the two schemes and their elements of intersection. The justification, such as it is, must be sought in a pragmatic appeal to the future" (S 31f). In his elaboration of the same point in *Process and Reality*, Whitehead admits that this is almost tantamount to saying that the "very meaning of truth is pragmatic" (PR 275), but he hastens to add that there must at some time be a definite determination of what is true for a particular occasion of experience because in the absence of such a determination the pragmatic test is unable to reach a judgmental decision and must resign itself to perpetual postponement.

What Whitehead seems to be saying is that the very possibility of applying the pragmatic test depends upon finally reaching a "day of judgment . . . when the 'meaning' is sufficiently distinct and relevant, as a perceptum in its proper mode, to afford comparison with the precipitate of feeling derived from symbolic reference" (PR 275f). When this situation is reached, it becomes possible to detect the agreement or disagreement between 'direct recognition' and the product of symbolic reference. If an error has been made, then it will be found "that some 'direct recognition' disagrees, in its report of the actual world, with the conscious recognition of the fused product resulting from symbolic reference" (S 19). 'Direct Recognition' is Whitehead's term for cognition which provides "immediate acquaintance with fact" (S 7) and it is a fundamental doctrine of the theory of symbolic reference that each of the components involved in any symbolic reference can be grasped in the mode of immediate acquaintance or direct recognition.

Although in any particular case one component, or set of components, functions as the symbol and is known immediately and directly, while the other functions as the meaning and is known mediately and indirectly, Whitehead does not distinguish between them on the basis of any intrinsic character. He does, however, say that "the more usual symbolic reference is from the less primitive component as symbol to the more primitive as meaning" (S 10) and also that "symbolism from sense-presentation to physical bodies is the most natural and widespread of all symbolic modes" (S 4). Thus, by this account, sense-presentations and physical bodies are equally subject to direct recognition and Whitehead feels that this avoidance of "any mysterious element in our experience which is merely meant, and thereby behind the veil of direct perception" (S 10) is also the avoidance of some metaphysical difficulties. A fundamental aim of the theory of symbolic reference is to overcome the difficulties which result from giving separate ontological status to sense-presentations and physical bodies, or to the mental and the physical. It is well known that Whitehead holds that there is a continuum of mind, body, and nature and opposes the Cartesian starting point of a dualism of the physical and objective as opposed to the mental and subjective. He attributes to medical physiology the credit for undermining this dualism and regards the chief issue of modern neurology, biophysics, and biochemistry as the replacement of the gap between the mental and the physical by a continuum. As he says, "It is a matter of pure convention as to which of our experiential activities we term mental and which physical" (S 20). It is largely from this basis that Whitehead argues that the difference between symbol and meaning is not substantial, but functional. It is only the function in a context of experience which marks a symbol and in another context the symbol could just as well function not as the symbol but as the symbolized meaning. He illustrates this by saying: "The word 'forest' may suggest of memories of forests; but equally the sight of a forest, or memories of forests, may suggest the word 'forest'" (PR 277). For a person reading a poem the word 'tree' may symbolize trees, but the poet seeking the words for his poem may regard the trees as symbols to suggest as their meaning the words he seeks (S 12).

The importance of this treatment of symbolic reference by Whitehead is revealed in his claim that "all human symbolism . . . is ultimately to be reduced to trains of this fundamental symbolic reference, trains which finally connect percepts in alternative modes of direct recognition" (S 7). But this statement is stronger than Whitehead intends, for in *Process and Reality* he admits that there can be symbolic reference between components from a single mode of direct recognition (PR 274). Although he regards the transference between alternative modes of perception as illustrating the fundamental principles of symbolism, whatever component occurs first is taken as the symbol component and the latter component is taken as the meaning. This degree of arbitrariness reveals the artificiality or conventionality of the symbol-meaning relation, but far from being merely a liability, it is this very aspect of indetermination which makes error and free imagination possible. If the error is a liability, it is also "the discipline which promotes imaginative freedom" (S 19). When the dog in Aesop's fable lost his meat in the attempt to grab its reflection in the water, he "made a mistake by reason of an erroneous symbolic reference" (S 20), "but he gained a step on the road towards a free imagination" (S 19). Now, while erroneous symbolic reference is possible and common, Whitehead regards it as impossible to have an erroneous direct recognition, or at least he says as much: "Direct experience is infallible. What you have experienced, you have experienced. But symbolism is very fallible, in the sense that it may induce actions feelings, emotions, and beliefs about things which are mere notions without that exemplification in the world which the symbolism leads us to presuppose" (S 6). Direct experience, or '' 'Direct recognition' is conscious recognition of a percept in a pure mode, devoid of symbolic reference" (S 19).

This last characterization of direct recognition raises a basic difficulty in Whitehead's account, because we have already seen that "perceptive experience, devoid of any symbolic reference, is in practice unobtainable" (S 54). On the one hand, he is asserting that all perceptive experience contains symbolic reference, and yet his very crucial thesis that it is symbolic reference which makes error possible, depends upon the detection of a discrepancy between the report of the world which is a product of symbolic reference and the report of the world given to direct recognition. In other words, the possibility of error may, as Whitehead maintains, depend upon symbolic reference, but the possibility of the detection of error depends upon a comparison of the deliverances of symbolic reference with the deliverances of an infallible perceptive experience which he calls direct recognition. This forces consideration of the question of how these apparently contradictory assertions might be reconciled.

But this is not the complete difficulty, because Whitehead makes a great deal of his contention that "symbolism . . . is the cause of progress, and *the cause of error*" [my italics] (S 59), or, as he says it somewhere else, "error is primarily the product of symbolic reference" (S 19). ". . . While the two pure perceptive modes are incapable of error, symbolic reference introduces this possibility" (PR 255). In spite of this, when he comes to treat the pure modes of perception, he says "In order to find obvious examples of the pure mode of causal efficacy we must have recourse to the viscera and to memory; and to find examples of the pure mode of presentational immediacy we must have recourse to so-called 'delusive' perceptions" (PR 186). It seems odd indeed that the chief examples of infallible perceptive experience should be visual delusions like double-vision, the feelings in 'phantom limbs,' deleria, etc. for the pure mode of presentational immediacy and notoriously fallible memory for the pure mode of causal efficacy. Perhaps the examples were poorly chosen, but in the total context of Whitehead's thought, one could hardly say they were not deliberately chosen.

If one is trying to make a consistent view out of the treatment of symbolic reference, he must, then, either abandon the claim that symbolic reference is the primary source of error, or he must abandon the claim to an infallible perceptive experience. We might consider the costs of the alternatives. Whitehead has himself set up three requirements for any epistemology, that it explain: "(i) how we can know truly, (ii) how we can err, and (iii) how we can critically distinguish truth from error" (S 7). He argues that his doctrine of direct recognition, or perception in the pure modes, answers requirement (i); that symbolic reference answers requirement (ii) "because it is only trustworthy by reason of its satisfaction of certain criteria provided by . . . [direct recognition]" (S 7). This makes the reliability of the deliverances of symbolic reference wholly dependent upon knowing their agreement with the deliverances of direct recognition. But if the infallibility of direct recognition is abandoned, this would entail the abandonment of reliable symbolic reference. Further, since, of the requirements for epistemology (iii) depends upon (ii) and (i), and (ii) depends upon (i) – at least if the error is conscious – the abandonment of the claim to an infallible perceptive experience would leave Whitehead's account without any explanation of how we can know truly, and the consequences of that. This does not seem to offer much hope as a means to resolve the contradictions.

Perhaps a more promising approach – though it raises difficulties of its own – is to be found in what may have been intended as Whitehead's answer to requirement (iii) for epistemology, that it explain "how we can critically distinguish truth from error" (S 7). Whitehead writes:

. . . Symbolic reference is a datum for thought in its analysis of experience. By trusting this datum, our conceptual scheme of the universe is in general logically coherent with itself, and is correspondent to the ultimate facts of the pure perceptive modes. But occasionally, either the coherence or the verification fails. We then revise our conceptual scheme so as to preserve the general trust in the symbolic reference, while relegating definite details of that reference to the category of errors. Such errors are termed 'delusive appearances.' This error arises from the extreme vagueness of the spatial and temporal perspectives in the case of perception in the pure mode of causal efficacy (S 54f). Our judgments on causal efficacy are almost inextricably warped by the acceptance of the symbolic reference between the two modes as the completion of our direct knowledge (S 54).

Whitehead elsewhere notes that, "No account of the uses of symbolism is complete without . . . recognition that the symbolic elements in life have a tendency to run wild" (S 61) and the passage just quoted at length above would seem to contain his advice on cutting back wild symbolism. Although he still speaks of 'the ultimate facts of the pure perceptive modes,' the passage does seem to contain the germ of a theory which could satisfy his three requirements without making the assumption that there are pure perceptive modes. Instead of regarding symbolic reference as trustworthy only if it satisfies the criteria of direct recognition, this passage suggests that symbolic reference be regardered as generally trustworthy until some 'wild' instance forces itself upon us. In this way, it is not necessary to regard the whole of symbolic reference as unreliable because of the encounter with errant instances. Rather, we respect the body of symbolic reference while being ready to revise details which fail to accord through verification or coherence. It is a fact that we are not deluded by appearances which we recognize as 'delusive.' In a discussion of the truth Whitehead writes: ". . . We can say that two objective contents are united in a truth-relation when they severally participate in the same pattern. Either illustrates what in part the other is. Thus they interpret each other. But if we ask what is meant by 'truth,' we can only answer that there is a truth-relation when two composite facts participate in the same pattern. Then knowledge about one of the facts involves knowledge about the other, so far as the truth-relation extends" (AI 310). Thus, wild symbolism is symbolism which fails to participate in the pattern of the composite fact of the present region. Perception in the mixed mode of symbolic reference "can be erroneous, in the sense that the feeling associates regions in the presented locus with inheritances from the past, which in fact have not been thus transmitted into the present regions" (PR 274).

Although the passage quoted above (S 54f) speaks of 'pure perceptive modes,' it seems to have abandoned the claim that these modes are infallible modes of perception since it specifically mentions "error . . . in the case of perception in the pure mode of causal efficacy". There is a passage in *Process and Reality* which associates the possibility of error with the other pure mode of perception: "This possibility of error is peculiarly evident in the case of that special class of physical feelings which belong to the mode of 'presentational immediacy'" (PR 390). On this basis, it seems possible to argue that we should drop the claim to infallible perceptive experience and also take seriously Whitehead's declaration that "complete ideal purity of perceptive experience . . . is . . . unobtainable" (S 54). Furthermore, if the distinct perceptive modes are fallible, then we need to qualify the assertion that symbolism is the cause of error although we might still admit that symbolic reference is the primary cause of error, especially inasmuch as we never actually have perceptive experience entirely devoid of symbolic reference. If, however, we cease to claim that direct recognition is infallible, or indeed, that it is absolutely direct, and if we cease to claim complete purity in the modes of perceptive experience, this need not lead us abolish the distinction between the two 'pure' modes of perception, i.e. presentational immediacy and causal efficacy, nor compel us to abolish the distinction between the 'pure' modes of perception and the 'mixed' mode of symbolic reference.

Because Whitehead's treatment of perception has been much discussed and is well known, I shall treat it only to the extent that its relevance to the theory of symbolic reference requires. That the theory of perception is intricately involved with symbolic reference is clear from Whitehead's statement

that "The conscious analysis of perception is primarily concerned with the analysis of the symbolic relationship between
the two perceptive modes" (S 81).

At the conclusion of the first Barbour-Page lecture at the
University of Virginia Whitehead says: "This lecture has
maintained the doctrine of a direct experience of an external
world" (S 28). In the course of the lecture he says something
(cited above) rather different: "Thus the result of symbolic
reference is what the actual world is for us, as that datum in
our experience productive of feelings, emotions, satisfactions,
actions, and finally as the topic for conscious recognition when
our mentality intervenes with its conceptual analysis" (S 18f).
Conceptual analysis is, after the two perceptive modes, the
third mode of experience. Conceptual analysis is said to
enhance the acuity of perception and to be present in all
conscious knowledge. In any case, if, with or without the
intervention of conceptual analysis, 'the result of symbolic
reference is what the actual world is for us,' it seems doubtful
that this could be summarized as defending the 'direct experience of an external world.' "When human experience is
in question, 'perception' almost always means 'perception in
the mixed mode of symbolic reference'" (PR 256). Whitehead
does not seem to have acknowledged this as a difficulty,
however, since he begins the second lecture with the words:
"It is the thesis of this work that human symbolism has its
origin in the symbolic interplay between two distinct modes
of direct perception of the external world. There are, in this
way, two sources of information about the external world,
closely connected but distinct. These modes do not repeat
each other; and there is a real diversity of information. Where
one is vague, the other is precise: where one is important,
the other is trivial" (S 30). If perception is almost always in
the mixed mode of symbolic reference, the actual world would
be the result of symbolic reference, but there could not be
'two distinct modes of direct perception of the external world.'
If, however, the 'two distinct modes of direct perception' be
regarded as analytic moments of actual perception in the
mixed mode of symbolic reference, then the contradiction
could be removed. I think that prior discussion and citation

provide evidence that this is a plausible interpretation of Whitehead's doctrine.

If, then, the distinct modes of perception are only analytically distinct, this reading will help to correct another confusion with regard to the two modes of perception. It is clear from passages too numerous to cite that Whitehead regards perception in the mode of causal efficacy as fundamental while perception in the mode of presentational immediacy is derivative. Although this is abundantly supported, one finds also indications that there is no basis for regarding one mode as more fundamental than the other. Whitehead tells us that "symbolic reference . . . is chiefly to be thought of as the elucidation of percepta in the mode of causal efficacy by the fluctuating intervention of percepta in the mode of presentational immediacy" but while this is chiefly the case, he tells us in the clause I just omitted that "in complex human experience it works both ways" (PR 271). Usually, primitive causal efficacy is symbolized by the less primitive presentational immediacy, but the reverse can be the case. Whitehead illustrates the functions of the two modes of perception with science as an example. Although scientific *observation* of all kinds seeks to restrict itself to presentational immediacy without any symbolic reference to causal efficacy, scientific *theory*, on the other hand, exerts just as great care to restrict itself to causal efficacy without any symbolic reference to presentational immediacy. As Whitehead puts it, "What we want to know about, from the point of view either of curiosity or of technology, chiefly resides in those aspects of the world disclosed in causal efficacy: but . . . what we can distinctly register is chiefly to be found among the percepta in the mode of presentational immediacy" (PR 257). Presentational immediacy, or the sense-perception which yields "clear, conscious discrimination, is an accident of human existence. It makes us human. But it does not make us exist. It is of the essence of our humanity. But it is an accident of our existence" (MT 158). As Whitehead describes it, presentational immediacy is "sense-presentation of the contemporary world" (S 80). Furthermore, "pure presentational immediacy refuses to be divided into delusions and not delusions"

(S 24). This being so, it is difficult to understand why White-head would say that this mode peculiarly evidences the possibility of error (PR 389). Its importance is severely restricted, ontologically at least, in that it concerns only a small minority of high-grade organisms. Presentational im-mediacy concerns the same datum as causal efficacy and this shared datum is one of the two elements of common structure which constitute the intersection needed for the possibility of symbolic reference. The other common structural element is the 'locality' or 'presented locus.' This locality is directly perceived in presentational immediacy; indirectly perceived in causal efficacy. "Presentational immediacy is an outgrowth from the complex datum implanted by causal efficacy" (PR 262). Together the datum and the presented locus unite the two modes of perception and form the basis for symbolic reference, "the synthetic activity whereby these two modes are fused into one perception" (S 18).

But besides the small minority of high-grade organisms capable of presentational immediacy, there are other organisms evidently not capable of this measure of differentiation and discrimination. Even in the absence of anything we should ordinarily call perception, we do find response in more primitive kinds of organism.

While sense-presentation is, in the main, an achievement of high-order organisms, Whitehead says that even "a tulip which turns to the light has probably the very minimum of sense-presentation" (S 4). This 'turning toward,' or re-sponse, is a primitive sort of perception and it is heavy with the sense of concealed powers. Low-grade organisms capa-ble only of perception in the mode of causal efficacy have especially, Whitehead says, "a sense for the fate from which they have emerged, and for the fate towards which they go" (S 44). The primitive emotions, "anger, hatred, fear, terror, attraction, love, hunger, eagerness, massive enjoyment, are. . . closely entwined with the primitive functioning of 'retreat from' and of 'expansion towards'" (S 45). But in hu-man experience, causal efficacy shows itself not so much in the primitive emotions as it does in what Whitehead calls the 'withness of the body.' "It is this withness that makes the

body the starting point for our knowledge of the circumambi-
ent world. We find here our direct knowledge of 'causal efficacy'" (PR 125). ". . . The animal body is the great central
ground underlying all symbolic reference. . . Every statement
about the geometrical relationships of physical bodies in the
world is ultimately referable to certain definite human bodies
as origins of reference" (PR 258f). All our perceptual knowledge stems from the bodily sense-presentations through the
sense organs. By a special effort of attention we can reflect
that we sense *with* our sense organs, but as we do so we are not
sensing the organs. If I hear *with* my ears, I do not hear my
ears. Thus, the 'withness of the body' is usually unconscious
but that does not discredit it as the source of the derivation
of conscious perception. Causal efficacy makes itself felt as a
force for conformation. In Whitehead's words, "Causal efficacy is the hand of the settled past in the formation of the
present" (S 50). He thinks that our notion of causation is abstracted from man's primitive perception in the mode of causal
efficacy, or conformation, and that human bodily experience
is chiefly "an experience of the dependence of presentational
immediacy upon causal efficacy" (PR 267).

Whitehead is quite conscious that his view contradicts
Hume. Hume, he thinks, was chiefly the victim of an abstract
notion of time as pure succession. "Time in the concrete,"
Whitehead affirms, "is the conformation of state to state, the
later to the earlier" (S 35). He thinks that not only common
sense, but also physics, and physiology are against Hume in
that they all recognize "a historic route of inheritance, from
actual occasion to succeeding actual occasion, first physically in the external environment, then physiologically --
through the eyes in the case of visual data -- up the nerves,
into the brain" (PR 260). Whitehead says that, generally
speaking, presentationally immediate perception finds its
expression in adjectives, while causally efficacious perception
expresses itself in substantives (PR 272). Besides the traditional priority of substance to attribute, this method of characterizing the relation of presentational immediacy to causal
efficacy as one of dependence lends further support to
the contention that causal efficacy is the aboriginal mode of

perception, contrary to the view of Hume. The distinctness, the definiteness, and the greater aptness for control and enjoyment of perceptions in presentational immediacy all contribute to making presentational immediacy "vivid, precise, and barren" (S 43). Although it is handy, it is trivial. Causal efficacy, on the other hand, is "vague, haunting, unmanageable" (S 43). "It produces percepta which are. . . not to be controlled, heavy with emotion: it produces the sense of derivation from an immediate past, and of passage to an immediate future; a sense of emotional feeling, belonging to oneself in the past, passing into oneself in the present, and passing from oneself in the present towards oneself in the future. . . . This is our general sense of existence, as one item among others, in an efficacious world" (PR 271). It is because most symbolization involves specific symbols but a rather vague meaning, that the symbolic reference from presentational immediacy as symbol to causal efficacy as meaning is Whitehead's choice to demonstrate the principles of symbolization. Also, since in the broader context, symbols tend to enhance the importance of the meanings they symbolize (S 63), symbolic expression, can, in the social context, lend strength to social stability by reinforcing instinct with emotion.

The instinctive reaction of an organism, or a society (and the two are interchangeable for Whitehead) is simply the primitive response of perception in the mode of causal efficacy. Once a high-grade organism becomes capable of symbolization, of symbolic reference, its action is no longer instinctive, but symbolically conditioned. "Symbolically conditioned action is. . . conditioned by the analysis of the perceptive mode of causal efficacy effected by symbolic transference from the perceptive mode of presentational immediacy" (S 80). In the absence of conscious effort and attention, an organism lapses from the activity of symbolic reference. Whitehead cites as an example of reflex action the disciplined person who responds to the sound of a command without having to work through the idea (S 75). Besides the symbols which produce superficial response, there are others, designed to instill a general sense of importance (S 75). The very survival of society depends upon sharing symbols which combine tactical directions with the more general strategic purposes of the society.

Man by his attainment of symbolic reference "can achieve miracles of sensitiveness to a distant environment, and to a problematic future. But. . . [he] pays the penalty, by reason of the dangerous fact that each symbolic transference may involve an arbitrary imputation of unsuitable characters" (S 87). Ultimate survival for any society depends upon finding the mean between reverence for the symbols which give the society stability and preparedness for revision which gives the society the adaptiveness to survive. Where there is no reverence for symbols, there is anarchy; where there is no willingness to revise them to meet new situations and needs, there is atrophy (S 88).

THE UNDERSTANDING OF THE PAST

RAMONA T. CORMIER

> If civilization is to survive, the expansion
> of understanding is a prime necessity.
>
> (MT 63)

In the contemporary world of competing ideologies and scientific advance geared toward human destruction, a progressive understanding of the past may lead to more prudent human action. If, however, historians are asked what this understanding entails, disagreement ensues. Understanding seems to fall into either one of two broad categories. It may mean either an interpretation of selected segments of the past by the ordering of relevant detail or a panoramic interpretation of the whole of mankind's past. I shall label the former scientific history and the latter universal history. If a historian accepts either of these as the nature of the historical enterprise, his attitude toward the other is usually determined. The scientific historian is apt to contend, on the basis of methodological principles, that universal history is not a legitimate study of the past. The universal historian, on the other hand, may reject the limitations set upon the historical enterprise by interest in minutiae and the specialization which results.

Contrary to these contentions, I shall maintain in this essay that the methods of these two types of history are not distinct. I shall justify the difference between them by arguing that each type of study is the consequence of the function of either practical or speculative Reason. Thus I shall endeavor to show that both species of history are legitimate intellectual enterprises essential to a progressive understanding of the past. These arguments will be set within the framework of Whitehead's cosmology and will furnish an ontological basis for both categories of history.

I

Studies of the past, like any intellectual endeavor, involve the selecting, ordering and interpreting of the material under examination. The controversy between the scientific and universal historian is waged over the legitimacy of the latter's principles of selecting and interpreting the data of the past. The scientific historian emphasizes the strict presentation of the facts. His concern, he may argue, is with the uniqueness of past occurrences and not with those characteristics held in common with other occurrences. From the point of view of the scientific historian, the works of the universal historian violate the principle of uniqueness. The sweeping generalizations which serve as the basis for the interpretations of the universal historian stress the common characteristics of past occurrences and not their uniqueness. I shall contend, in disagreement with the scientific historian, that his methods and those of the universal historian do not differ in kind but only in degree of generality and abstraction. I shall support this position by arguing in this section of the essay first that any occurrence is at the same time unique and similar to other occurrences and second that relations between occurrences are designated by common characteristics determined by certain methodological factors delineating the form and content of a narrative. Whitehead's analysis of the actual occasion will serve as the ontological justification of the first contention.

The historian, whether scientific or universal, believes that his account of the past is about what actually happened. What actually happened, according to Whitehead, is the actual world built up of actual occasions (PR 113). It is from actual occasions that "whatever things there are in any sense of 'existence,' are derived by abstraction" (PR 113). The actual occasion refers to the extensiveness of an actual entity (PR 119). This spatio-temporal extensiveness of an actual occasion has three characters, e.g., it is separative, prehensive and modal (SMW 94). The nature of these aspects of the actual occasion determine its uniqueness and its relations to other actual occasions. The separative and modal characters designate its uniqueness; the prehensive character, its relatedness.

The spatio-temporal extension of an occasion is separate from other occasions. This separateness of the occasion is definitely determined by its spatial or temporal modality. Spatial modality designates the sense in which an occasion is at this place and no other. Temporal modality is the endurance of an occasion during a certain period and through no other (SMW 94). The individuality of an occasion resides in its separative and modal characters. One could argue that an occasion has a unique set of prehensions; i.e., no two occasions are related to the same collection of occasions, nor if they were related to the same collection, would the relations be the same. This unique character of an occasion's prehensions, however, depends upon its separative and modal aspects, i.e., an occasion has a certain set of prehended occasions because of its peculiar spatio-temporal extensiveness.

The unique character of an actual occasion is not a sufficient basis for a history. A history is not a series of occasions chronologically listed but a collection of selected occasions ordered and interpreted according to a particular point of view. It is the prehensive character of an actual occasion or its spatio-temporal togetherness with other occasions which ontologically justifies the relation of occasion to occasion in a history. "'Together' is a generic term covering the various special ways in which various sorts of entities are 'together' in any one actual occasion" (PR 32). This togetherness is a process of unification whereby each occasion is something from the standpoint of every other and also, from the standpoint of every other, is something in relation to it (SMW 101–102). The process of unification is an uncognitive apprehension, i.e., an apprehension which may or may not be cognitive (SMW 101). It is a process consisting of three factors: the "subject" prehending, the "datum" prehended and the "subjective form" which is how the subject prehends the datum (PR 35). A rock may be the subject of a prehension, a drip of water, the datum, and the conditioning of the surface of the rock by the water, the effect of the datum on the subject. This is an example of a simple physical prehension or an act of causation (PR 361). My prehension of the red patch on my desk as a book is a more complex prehension involving the integration of a physical and conceptual feeling (prehension). A conceptual feeling is a feeling whose datum is an eternal object (PR 367). In the latter case "book" is the conceptual feeling integrated with the sensation "red patch." All a-wareness involves the synthesis of physical and conceptual prehensions (PR 372). Thus the subject-matter of history is grasped through the synthesis of physical and conceptual prehensions.

Although actual occasions ontologically justify the subject-matter of history, what the historian grasps are prehended aspects of actual occasions. "A prehensive occasion is the most concrete finite entity, conceived as what it is in itself and for itself and not as from its aspect in the essence of another such occasion" (SMW 104–105). A prehended occasion is grasped as a propositional feeling. A propositional prehension is an integrated synthesis of a physical feeling with a conceptual feeling (PR 393). The primitive level of propositional prehensions identifies prehended occasions spatially or temporally. The historian's account is often of occasions he has not himself experienced. Hence the physical aspect of the propositional prehension is either memory (if the occasion was witnessed), evidence or traces of the past in the present. From these sources the historian infers that occasion y occurred at either time t or place x. (a) "On the 16th Napoleon left Paris" is an example of such an identification.[1] Any historian begins his selection from an indefinite number of such propositions to be termed hereafter "historical facts."

The confirmation of the simple historical fact is a complicated method whereby evidence or traces of the past in the present are analyzed and authenticated. This essay is not concerned with the methods of confirming historical facts. The dimension of historical fact, i.e., the level of simple facts which have been confirmed, is mentioned because it is the starting point of any study of the past and is that aspect of historical knowledge agreed upon by most historians. If there is disagreement as to the credibility of a historical fact, it is

---

[1] Julius von Pflugh-Harttung, "The War of Liberation 1813–14," in *The Cambridge Modern History* (New York, 1906), IX, 224.

a consequence of lack of evidence or of previous knowledge. Disagreement would be justified in such a case by the inadequate grounds for confirming the fact.

From the primitive level of historical fact, the historian may formulate one or more types of complex historical facts of a higher level of generality and abstraction. The general proposition so formulated ultimately rests upon the particular primitive historical facts which are its justification. One such type of fact refers to an aggregate of prehended occasions which may have the approximate time, place or type of prehended occasion in common. "Finally Shakespeare and Cervantes died on the same day, April 23, 1616" is an example of that type of complex historical fact referring to two or more primitive facts identifying occasions (SMW 58). The occasions of the deaths of Shakespeare and Cervantes are integrated in a complex fact by Whitehead in his effort to show the coincidences which marked the literary annals of the seventeenth century, that century of genius which furnished the ideas upon which the past two or more centuries have been living (SMW 57–58).

There is a more complex historical fact of a higher level of generality and abstraction which is illustrated by the following quotation from *Adventures of Ideas:* (c) "Throughout the Hellenic and Hellenistic Roman civilizations – those civilizations which we term 'classical' – it was universally assumed that a large slave population was required to perform services which were unworthy to engage the activities of a fully civilized man" (AI 14). The spatio-temporal slab referred to in this case is much more comprehensive than that of either (a) or (b). In the case of (c) "universally assumed" is relevant to the idea of slavery in the political writings of the Greeks and Romans. Before (c) was formulated and given significance in a history, the foundation for the abstraction of the concept of slavery from the political writings of Romans and Greeks had to be determined by fundamental principles of a high degree of abstraction and generality. Whitehead has stated that the intellectual agencies involved in the modification of epochs are the proper subject of an adventure of ideas (AI 19). These intellectual agencies may be either general ideas or highly specialized notions. The concept of slavery in the political writings of the Greeks and Romans is an example of such a general notion. How this concept was modified and replaced by the concept of freedom can be interpreted as Whitehead's attempt to apply to a particular period of the past his cosmological theory of progress which rests upon the modification or replacement of ideas which form the cosmological outlook of an epoch by other ideas.

The primitive level of historical fact conceives of the prehended occasion "in itself and for itself," e.g., in its uniqueness. Complex historical facts are unique in the sense that they refer to certain primitive historical facts and to no others. Yet the particular primitive facts related in a complex fact are integrated by their common characteristics. Complex historical facts of varying degrees of generality and abstraction may occur in scientific or universal history. The more general proposition at the cosmological level, however, is more apt to occur in a universal history. The type of complex fact and the selected primitive facts to be so integrated are determined by basic principles of interpretation.

It is in the ordering and interpreting of selected simple or complex facts that rank disagreement among historians arises. The structure of any ordered interpretation of the past depends upon the "togetherness" of actual occasions. The relation of fact to fact is justified by common characteristics. The particular characteristics of selected facts to be signalled out is designated by methodological factors basic to any history. Certain aspects of a historian's spatio-temporal perspective specify the general frame within which a scientific or universal history is constructed. These methodological factors are (1) the historian's definition of the subject-matter of history, (2) the extension of our knowledge and beliefs about man's social behavior as we relate more of the unknown to the known and (3) the alteration of the historian's views of what is "significant" in the past as his understanding of the consequences of what has happened changes with what comes to pass.[1]

[1] See John Herman Randall, Jr., *Nature and Historical Experience* (New York, 1958), pp. 40–41.

(1) What the historian determines as the subject-matter of history will limit the dimension of historical fact relevant to his studies of the past. This restriction may be set by the dictum "how it actually happened" and within the confines of this dictum one may designate certain aspects of the dimension of historical fact as relevant to studies of the past, e.g., one may confine history to human affairs and within this cadre, to past politics.[1] Or one may designate the whole of the known past as the periphery of one's account and within this periphery specify man's spiritual affairs as the subject-matter of history.[2] In each case the level of generality and abstraction will differ in degree according to the spatio-temporal extensiveness to be covered. If the strict presentation of the facts is one of the defining characteristics of history, then the historian's narrative is confined to description and to low level explanation. If a panoramic vision of the past is the historian's intention, then his account will be a correlation of simple or complex facts of varying degrees of generality and abstraction.

(2) Methodologically speaking, the available primitive facts and the possible relations between them depend upon the spatio-temporal location of the historian. As man relates more of the unknown to the known, the dimension of historical fact may be extended or systematic knowledge of types of behavior may be enlarged. Discoveries of hitherto unknown evidence may alter the dimension of historical fact. Archeological findings may furnish the evidence or traces of the past which lead to the filling in of gaps or to the modification of confirmed facts. In addition man's increasing knowledge about himself and his environment enables him to give more than one explanation of an occasion. Thus, he may render a biological, psychological, physiological, economic or other explanation of a human relation. Systematic areas of intellectual endeavor are not static activities. They are subject to modification.

(3) The historian's views of what is significant change with

---

[1] See E. A. Freeman, *The Methods of Historical Study* (London, 1886), p. 44.
[2] See Arnold J.Toynbee, *A Study of History* (London, 1954), IX, pp. 168–69.

his understanding of the consequences of what has happened in the light of what comes to pass. The dimension of historical fact is not a closed context. It is continuously being added to as occasions become actualized. As the dimension of historical fact changes, so does the historian's own dimension of historical fact become extended as his present experiences move into the past. Hence his views affecting his methods of selecting, ordering and interpreting may change as his understanding and experience evolve.

In addition to these factors, the historian employs other methodological procedures which determine more specifically the form and content of his study. These latter factors are (a) the context of a study of the past, (b) the perspective of the context, (c) the point of view of the perspective and (d) the specific problem under consideration.

(a) The spatio-temporal segment of the dimension of historical fact the historian chooses to relate is the context of his narrative. For example, the context may be 18th century France or France from its inception up to the present. It would be impossible for any historian to construct a study of the past which would include all of the extant historical facts in any context. The other factors in this group specify the facts to be selected and interpreted within the confines of what is designated as the subject-matter of history.

(b) Within the context, the historian selects a perspective for his account of the past. A historian whose context is 18th century France may choose any of the following as his perspective: ideas, architecture, the French Revolution, Napoleon or others. The perspective is the sub-ordering of facts within the context designating a unifying factor of the account, As the spatio-temporal range of the context increases the number of possible perspectives also increases.

(c) The perspective selected may be interpreted from one of several points of view. Within the perspective the point of view determines the relationships between facts. The specific facts and the mode of relationship between them is designated by either principles, generalizations, laws, normative statements or other factors. It is the point of view which sets the conditions for an occasion, thus determining the specific facts from the perspective or other related relevant sub-orderings of facts to be interpreted.

(d) The context, the perspective and the point of view form the structure within which the historian attempts to resolve a specific problem. The problem may take the form of the tracing of a tendency, the causes of a war or revolution, and so on. The problem refers to what it is the historian is trying to show or resolve by certain relationships between facts.

Many different orderings and interpretations of historical facts are possible. The comprehensiveness of the context increases the number of perspectives possible. Within a single narrative, facts from several different perspectives may be relevant to the account. It is the task of the historian to weave a consistent and credible narrative of the historical facts relevant to the problem under consideration. As the spatio-temporal compass of the context increases, these possible relationships increase. The larger the segment, the more general and abstract is the account.

Whether a history falls under the scientific or universal category, the methods employed in relating facts to other facts are primarily the same. As I have pointed out, the level of generality and abstraction differs. At either level the historian is attempting to establish relationships between historical facts which will extend our understanding of the past. Understanding is achieved by increasing the number of justified connections between facts. The scientific historian is not willing to accept the pattern or scheme whereby the universal historian justifies his relations because of its generality and abstraction, yet he is willing to accept some connections and to increase thereby his degree of penetration. Since he is not willing to offer a methodological justification for universal history, I shall seek an ontological justification in the distinction Whitehead has made between the practical and speculative Reason and the modes of understanding which evolve.

II

According to Whitehead, understanding has two modes of

advance, e.g., "the gathering of detail within assigned pattern and the discovery of novel pattern with its emphasis on novel detail" (MT 80). These two modes of understanding are a consequence of the function of the speculative and practical Reason. These functions of Reason are essentially distinguished by the purposes governing its operation in each case. The operations of practical Reason, according to Whitehead, are motivated by interests external to itself. Speculative Reason serves only itself. This distinction in motivation determines the intellectual activity which proceeds from the operation of these two aspects of Reason. Practical Reason is responsible for "the piecemeal discovery and clarification of methodologies," speculative Reason "seeks with disinterested curiosity an understanding of the world" (FR 29). Thus, practical Reason nurtures the methodologies of every special area of intellectual activity, while speculative Reason fosters cosmology.

Practical Reason is bound by the limits of a successful method. These limits lead to Obscurantism, that state of inertial resistance arising when recent habits of speculation interfere with a fixed method (FR 34). In this respect, practical Reason is opposed to speculative Reason. Speculative Reason is continuously engaged in attempts to transcend any particular method. Its attempts at transcendence take either one of two forms: the transcendence achieved within the boundaries of a fixed method or the transcendence achieved through the construction of a most general interpretative scheme of the present stage of the universe. In the former activity, speculative Reason is operating in alliance with practical Reason and is endeavoring "to enlarge and recast the categoreal ideas of a particular methodology within the limits of that topic" (FR 68). In the latter, speculative Reason is engaged in the construction of cosmologies. It is practical Reason which gathers the detail within an assigned pattern and in so doing advances understanding. Speculative Reason, on the other hand, discovers novel patterns emphasizing novel detail.

Neither speculative nor practical Reason can conduct its operations apart from the other. Speculative Reason discovers

the categoreal schemes of each specific method. It also en-
larges and recasts these ideas when the need arises. In its
most comprehensive function, speculative Reason is engaged
in the discovery of a most general interpretative scheme which
is presupposed by specific methodologies at specific times.
Speculative Reason is protected from illegal flights of fancy
by practical Reason. The novel concept must accord with a
specific methodology. The discovery of a novel concept leads
to disagreement between a cosmological outlook, a special
science and the novel concept. Practical Reason and speculative Reason function together in the modification of each
of these endeavors. Practical Reason collects and examines
the relevant evidence and speculative Reason formulates the
modifications of all three so that accord results (FR 70).

Like all efforts of finite intelligence, cosmologies fail to
achieve the generality and clarity at which they aim. Cosmology is a formulation of the most general interpretative
scheme of the present stage of the universe. It generalizes
beyond any special science, and thus provides the interpretative system which expresses their interconnection. "Cosmology, since it is the outcome of the highest generality of
speculation, is the critic of all speculation inferior to itself in
generality" (FR 69).

How do scientific and universal history fit into Whitehead's
analysis of Reason? The scientific historian is engaged in the
elaboration of detail within fixed methods. The scientific
historian constructs his narratives within the confines of
specific methods. He orders and interprets facts within accepted perspectives and relates them according to generalizations and laws which form the structure of special methods.
In the case of the universal historian the form of his study of
the past is not confined to that of specific methods. He is
attempting to construct a scheme which will relate the
various specific methods to each other. In so doing, he engages
in the discovery of novel concepts and patterns for interpreting the past. If the universal historian endeavors to
construct a most general interpretative scheme, then, according to Whitehead's analysis, his universal history is a
cosmology. His conceptual scheme is cosmological in gener-

ality and purpose, but his universal history is not simply the formulation of such a scheme but is an endeavor to interpret the past according to the dictates of the scheme. Thus the universal historian not only invents a novel cosmological scheme but also attempts to verify his pattern by his study of the past. If Whitehead's criterion of an adequate cosmology is applied to universal histories, then one discovers that many of the schemes recast and enlarge one special method and use it as the basis for the most general interpretative scheme. The Marxian historian reduces human activity to the economic level; Toynbee gives precedence to religious concepts and principles. Cries of unwarranted relations are based upon this failure of the universal historian to find a most general scheme transcending a specific method.

The universal historian has attempted to come to grips with the general scheme presupposed by all special methods. His scheme will sooner or later become inadequate, but in the process of his construction of a study of the past, he has discovered relations and increased the depth of understanding. His novel conceptual scheme is kept within bounds by the attempt to interpret the past according to the scheme. Thus practical Reason is performing its function in the construction of a universal history. Universal histories, on the other hand, might lead to the modification of special methods. The distinction between the two categories of history is justified by Whitehead's analysis of the practical and speculative Reason. The operations of both are essential to the historical enterprise and to the advance of human understanding.

CAUSAL EFFICACY AND CONTINUITY IN WHITEHEAD'S PHILOSOPHY

HAROLD N. LEE

Without the perception of causal efficacy, two essential parts of Whitehead's philosophy fall apart. He maintains that his cosmology is realistic. He also declares that his philosophical outlook is basically empirical. The perception of causal efficacy is the factor that holds these two things, the realism and the empiricism, together. The basis of his realism is his doctrine of the extensive continuum, but if all empirical knowledge is to be traced to what Whitehead calls "presentational immediacy," no knowledge of the extensive continuum can be obtained, and thus no knowledge of the actual world in which that type of high grade organism called "mind" is immersed and of which it is a part.

Presentational immediacy is that which is given in sense perception; i.e., qualia – Hume's impressions – and the aggregates of qualia which have simple spatial structure. Presentational immediacy does not yield temporal connections; it reaches neither toward the future nor into the past. An appeal to memory does not yield the past, for memory is given in the present. Presentational immediacy does not reach into the future because without reference to the past there is no justifiable expectation. Even when there is expectation, the imagery in which it is embodied exists in the present.

If, nevertheless, what is given in presentational immediacy is said to be in time because of its constantly changing character, the time involved is only pure succession, in which one instant or thing or character is replaced by another, and this is not continuous. Whitehead has no use for the doctrine of time as pure succession (S 44–45, 34–35). Time is continuous only if the past reaches into the present and the present into the future without absolute dividing lines.

I

Hume began with sense impressions. On this foundation
he erected an empirical epistemology, but it will not yield a
cosmology. Each sense impression is complete in itself, and
the impressions merely follow each other in an otherwise
empty time. Thus, Hume's time is pure succession (S 43).
Cosmology, however, is a theory of the process of becoming,
and requires temporal connectedness of the world, not a mere
succession of states; and if the cosmology is to be empirical,
the connectedness must manifest itself in perception. No
such connectedness is to be found in the data of presentational immediacy. One cannot get change by stringing together unchangnig things or states. Whitehead says that
presentational immediacy is vivid and precise but barren. It
displays what is bound together only "in an impartial system
of spatial extension" (S 23). Things that are given in presentational immediacy "happen in complete independence except
for their spatial relations at the moment" (S 25). Whitehead
refers with approval to Santayana's reduction of any such
theory as Hume's to "solipsism of the present moment" (S 28,
33).

Whitehead maintains, then, that there is perceptual experience of temporal connectedness, and he adduces evidence
that there is (S 41–49). The doctrine of perception of causal
efficacy must be reasonably justified by empirical evidence.
If it is an *ad hoc* hypothesis, it will not support the empiricism
of the cosmology. To hold that there is perception of temporal
connectedness requires that relations as well as qualia be
directly perceived (S 43). The perceived relations are spatio-
temporal, and the perception of them is the ground of our
knowledge of the extensive continuum.

The doctrine of the perception of causal efficacy rests on
the experience of the past reaching into the present. The
disappearing portion of the specious present is perceived
to be immediately relevant to what it becomes. "Causal efficacy is the hand of the settled past in the formation of the
present" (S 50). Causal efficacy is the conformation of the

present to the immediate past, and both the immediate past and its relevance to the present are experienced. Experience does not yield a knife-thin edge between a past which is gone and a future which is not. What we experience is a specious present that is "thick," thick enough to contain the immediate past which is not yet gone but is in the process of perishing, and thick enough to contain the future which has not yet become but is in the process of becoming. The present is a duration. There is no absolutely instantaneous experience (CN 57). Instantaneousness is a concept derived by extensive abstraction. Experience, on the other hand, is concrete.

If we think of perception taking place in an instantaneous present, and the object of perception existing in an instantaneous present, we have made the assumption of simple location. Both Hume and Kant made this assumption (S 38). It is compatible with the Newtonian doctrine of absolute space and absoute time but is not compatible with the principle of relativity. The assumption of simple location, by abstracting from duration and discarding the relevance of the past in the present and the conformation of the present to the past, gives rise to the fallacy of misplaced concreteness (S 39). Conscious concrete experience is the awareness of the present in its relation of conformationto the actual immediate past.

In a high grade organism, the awareness of the conformation of the present to the immediate past is always conjoined with the data given in presentational immediacy. The relationship between the two perceptive modes – the perception of causal efficacy and the perception of presentational immediacy – whereby the one finds meaning in terms of the other is what Whitehead calls "symbolic reference" (S 18). "Complete ideal purity of perceptive experience, devoid of any symbolic reference, is in practice unobtainable for either perceptive mode" (S 54). Thus, the distinction between the two modes is a distinction found by analysis, but it is *experience* that is analyzed. Whitehead is working consistently within an empirical frame of reference.

Hume argued that there is no perception of the relation between cause and effect because: 1) he assumed that the only data of perception are the discrete, ready-made units of sense impressions which have no connection other than that they succeed each other; and 2) because he was looking for a highly sophisticated, conceptual relation that is a function of symbolic reference instead of looking for a primitive feeling of relevance of the immediate past to the present and the conformation of the present to the immediate past. Whitehead argues that such a feeling is part of the data of experience. If it is, a realistic and empirical cosmology is possible.

II

The foregoing account of Whitehead's doctrine of the perception of causal efficacy and of presentational immediacy and of the interaction of the two in symbolic reference is based on his *Symbolism, Its Meaning and Effect*. This was published in 1927, between *Science and the Modern World* (1925) and *Process and Reality* (1929). The doctrine of the perception of causal efficacy is fundamental in *Process and and Reality*, and its elaboration in *Symbolism, Its Meaning and Effect* must be considered as a preparatory study. It is primarily an expression of the underlying importance of continuity for Whitehead. Tne flux of experience is the point of departure for all empirical philosophies.

Emphasis on the flux of experience is present throughout Whitehead's philosophic writing. He points out in *The Organization of Thought* (1917) that the experienced present is a duration in which there is past, present and future (OT 110, 145). ". . . the flux of things is one ultimate generalization around which we must weave our philosophical system" (PR 317). He says in *The Function of Reason* that "no notion could be further from the truth" than that "conscious experience is a clear-cut knowledge of clear-cut items with clear-cut connections with each other" (FR 62).

Whitehead is surely correct in holding that conscious experience is not a set of discrete sensations. The most profound difficulties in Locke's and Hume's empiricism stem from their assumption that it is. There are no *experienced* ultimate disconnections. Perception in the mode of causal efficacy is the acknowledgement of the spatio-*temporal* continuity of experience. Unless continuity is ontologically fundamental, there can be no process cosmology.

In emphasizing the continuity of nature, Whitehead says, "No individual subject can have independent reality, since it is a prehension of limited aspects of subjects other than itself" (SMW 217). He is here talking primarily within the context of the "subject-object relation," but he explicitly rejects the Cartesian bifurcation, and "subject" is not limited to mind. What he says about the subject in the sentence quoted applies to all actual entities in *Process and Reality*. He remarks that his point in this passage is that we must start from "the analysis of process as the realization of events disposed in an interlocked community" (SMW 219).

III

The cosmology elaborated in *Process and Reality* is a theory of the becoming of actual entities. Concrescence is the coming into being of actual entities through prehension of other actual entities and the ingression of eternal objects. "The actual world is a process, and the process is the becoming of actual entities" (PR 33). The actual entities are atomic. The philosophy of organism is "an atomic theory of actuality" (PR 40). Note here that he says "actuality" not "reality." To Whitehead, potentialities are real but are not atomic (PR 103–104).

The atomism of *Process and Reality* is not easily compatible with the fundamental continuity necessary to his doctrine of the perception of causal efficacy. Yet, if the continuity is not there, the perception of causal efficacy cannot be there; and if the perception of causal efficacy is not there, Whitehead's cosmology lacks an essential part of its empirical support.

"So far as the contemporary world is divided by actual entities, it is not a continuum but is atomic" (PR 96). The actual entities, it is emphasized again and again, are atomic. "The ultimate metaphysical truth is atomism" (PR 53). When Whitehead is in this mood, atomism seems to be more fundamental than continuity and seems to be contrasted with continuity. Although there is creation of continuity in the present cosmic epoch, it "does not seem to be a necessary conclusion" that continuity is an ultimate metaphysical truth (PR 53).

IV

It is the contention of the present essay that the atomism expressed in *Process and Reality* is not necessary to White-head's system or to any empirical cosmology. Furthermore, it is an intrusion into Whitehead's system; for if the atoms are discrete and discontinuous, the cosmology collapses; while if they are not discrete and discontinuous, the term "atom" is being used in a Pickwickian sense; and they cannot be both continuous and discontinuous as these terms are contradictory.

The atomism of Whitehead's cosmology first appears fully developed in *Process and Reality*, though it is adumbrated in *Science and the Modern World*. The actual entities are the atoms, and the term "actual entity" plays an important role for the first time in *Process and Reality*. In earlier works, the corresponding term was "actual occasion" or "actual event." On pages 27 and 32 of *Process and Reality*, he seems to say that "actual entity" and "actual occasion" denote the same thing, but the actual occasions of his earlier writing are not defined in those writings to be atomic whereas actual entities are atomic.

The foreshadowing of atomism in *Science and the Modern World* is to be found in the atomic nature of time in the epochal doctrine of time. Here, Whitehead is not using "time" as he uses it when denying, in *Symbolism, Its Meaning and Effect*, that time is pure succession. In *Science and the Modern World*, he distinguished between durations and temporalizations. Durations are the time of temporal experience (CN 56–59). Temporalization yields time that "is atomic (i.e., epochal) though what is temporalized is divisible" (SMW 185). The atomic time is "spatialized." [1] This yields the value for

---

[1] Whitehead often acknowledges his indebtedness to Bergson (CN 54), though he does not accept the doctrine of the élan vital, and he explicitly rejects Bergson's anti-intellectualism (SMW 74).

t in the equations of physics. Time as experienced is duration, it is continuous; but the time of physics is atomic, that is epochal. An epoch is an arrest. The epoch needed by the historian is a period of time bounded by limits, and a limit in time is an arrest.

It is not altogether clear whether Whitehead intends the atoms in his system to be discontinuous. If one looks at his system as a whole with its emphasis on organism, internal relations and prehensions, it seems obvious that there is no ultimate discontinuity in it. Yet, why is there such a heavy emphasis on atomicity; and why does he say (as quoted above), "So far as the contemporary world is divided by actual entities, it is not a continuum"; especially when "actual entities. . . are the final real things of which the world is made up" (PR 27)? "Continuity concerns what is potential; whereas actuality is incurably atomic" (PR 95). On page 470 of *Process and Reality*, he allows, at least hypothetically, for the individual discreteness of actualities.

On the other hand, there is an abundance of passages that would seem to indicate that the actual entities are not discontinuous. There could be no historic route of actual entities if they were discrete. "The actual occasions. . . constitute a continuously extensive world" (PR 105). "Cosmology must do equal justice to atomism, to continuity, to causation, to memory, to perception, to qualitative and quantitative forms of energy, and to extension" (PR 365); and it is to be noted that these different terms are not presented as being incompatible with each other. Furthermore, actual entities may be divided into prehensions (PR 28), and he says, "I use the term 'prehension' for the general way in which the occasion of experience can include, as part of its own essence, any other entity" (AI 300). Thus, there is evidence that Whitehead does not hold the atomicity of actual entities to be discrete.

The questions still remain: 1) exactly what did Whitehead mean by calling actual entities atomic; 2) why did he insist on atomicity? He cannot be using the word "atom" in the original sense, for his atoms are not simple and indivisible. Actual entities are made up of prehensions, and through prehensions, they are interpenetrating. The interpenetration is *essential* to the actual entity being what it is, and this is a theory of continuity, not discreteness. The view that the actual entities are ultimately separate would be a manifestation of the fallacy of misplaced concreteness.

"An actual occasion is analysable. The analysis discloses operations transforming entities which are individually alien, into components of a complex which is concretely one" (PR 322). "An actual occasion is nothing but the unity to be ascribed to a particular instance of concrescence" (PR 323). These quotations give, I think, the kernel of what Whitehead means by the atomic nature of the actual entity, and they do not indicate any fundamental discreteness. The actual entity is a unity; a unity of the sort we mean when we speak of individuality. "As used here the words 'individual' and 'atom' have the same meaning, that they apply to composite things with an absolute reality which their components lack" (AI 227). The only sense in which an actual entity is indivisible is that it cannot be divided into other actual entities (PR 334). Prehensions, on the other hand, are not atomic, because "they can be divided into other prehensions and combined into other prehensions" (PR 359).

I suggest that there are two main reasons for the emphasis on atomicity in *Process and Reality* and *Adventures of Ideas*. One is Whitehead's insistence on the absolute determinateness of the past, or the definiteness and division which is characteristic of actuality. The other is an endeavor to do justice to "the atomism of the modern quantum theory" (PR 365).

The difference between potentiality and actuality to Whitehead is that potentiality is whatever may be divided and actuality is that which is being or has been divided. Actuality is that which is definite; it is exactly what it is. "Actual entities perish, but do not change; they are what they are" (PR 52). Process no longer takes place in the past. The past has perished and has passed into objective immortality. The process is concrescence, the becoming of actuality, and this is the dividing of the potentiality of the extensive continuum by means of physical prehensions, both positive and negative, and by means of conceptual prehensions, which are the ingression of the pure potentiality of eternal objects. The division, to Whitehead, breaks the continuum into units, and the units are the atoms, the actual entities. A paradox arises here: no actual entity is *completely* definite until it perishes (but I concede that the appearance of the paradox may be due to my inability to interpret Whitehead here and not due to Whitehead himself).

It may be that the earlier forms of the quantum theory were interpreted in the light of absolute units of physical energy, but at the very time when Whitehead was writing *Process and Reality*, Schrödinger, Heisenberg, Dirac, de Broglie, Born and others were engaged in the researches that led away from the emphasis on atomicity. The present form of the quantum theory cannot be said categorically to demand the view that energy transformations are discontinuous.[1]

Whether or not Whitehead intended it, atomism brings with it the connotation of discreteness, and this connotation introduces confusion (or at least difficulty of interpretation) that is the source of apparent paradoxes in his philosophy. If the actual entities are discrete, the cosmology breaks down, for it is based on a primary ontological continuity.

V

If Whitehead had not used the term "actual entity," but had staid with his other term "actual occasion," there would have been less danger of confusion about the place of discreteness in his system. He says that "actual occasion" is used to emphasize the extensiveness of the actual entity (PR 119). Extensiveness is continuous. Continuity, to Whitehead, implies the possibility of division, and divisions once made are irrevocable. This is what he means by "decision." The determinateness of actuality is due primarily to the prehension of eternal objects (conceptual prehension), which are pure potentials determining the definiteness of actual entities; they have no actuality apart from their ingression into the becoming of actual entities (PR 34).

[1] See A. Landé, *Foundations of Quantum Theory: A Study in Continuity and Symmetry* (New Haven, 1955) especially at p. 83, "Are There Quantum Jumps?"

Whitehead realizes that division and decision involve selection. This is the reason for the doctrine of negative prehension. I suggest that although selections from the continuum that have been made *are* irrevocable, not all possible selections have been made. The possible selections from a continuum are infinite in number. This follows from his recognition that a continuum is infinitely divisible. Thus, the irrevocability of the past is not absolute, and, in fact, a new historical interpretation or any reinterpretation of the past involves new selections. The old selections are not obliterated, and thus the past is not altered; but new selections are added to those that have already been made, and the way that the past impinges upon the present can change, not physically but conceptually. There is no compelling reason to hold that actuality, in its perishing, has crystallized into indivisible and unalterable absolute units. The past is a part of the extensive continuum, and is continuous with the present which *is* in process. There is nothing either in the meaning of actuality or in actuality as we find it that requires absolute units in either the present or the past. Whitehead may hold that there is, but he does not show that within his system it is necessary so to hold.

There is a widespread covert assumption in philosophic thought that the process of analysis, if rigorously carried out, leads to units which are further unanalysable. Whitehead seems to make this assumption in so far as actuality is concerned, and arrives at units of actuality which are not further divisible into other actualities, and these units are the atoms, the actual entities of *Process and Reality*. It is a contention of the present essay that there is no warrant for this assumption in any evidence adduced either in experience or in mathematics. All that empirical evidence shows is that we *do* stop our analysis at some point, and that this may be the end of our present rope; but the history of science shows that whenever we have thought we have reached the ultimate indivisibility, we have been wrong. Mathematics never reaches an ultimate unit when dealing with either a dense series or a linear continuum. All that mathematics needs is a cut in the continuum, and a continuum can be cut anywhere.

VI

Let us suppose that the indivisible unit of actuality is not a necessary notion. Then Whitehead could have taken as his unit not the actual entity but what he calls the historic route of actual entities, and many of the puzzles that adhere to the attempt to understand his cosmology would disappear. The actual occasion, in this amended view, is a fragment of a historic route, but it is a fragment whose limits are not fixed. Each actual occasion is itself a continuum, and these continua are themselves parts of a more extensive continuum. The perception of causal efficacy is the empirical evidence that actual occasions interpenetrate. The doctrine of prehension is the statement of the way that each actual occasion is composed of other actual occasions and the ingression of eternal objects. The difference between an actual occasion and a historic route is a difference only of degree of extensiveness. Each actual occasion is an individual, and the historic route of that actual occasion is also an individual of greater extensiveness. Within the historic route, divisions are made, but these divisions are not an absolute property of the historic route. They are made by the ingression of eternal objects, which are conceptual prehensions. Every actual occasion and thus every historic route is an individual that prehends and can be prehended. The prehension is a process. The "can be prehended" is in the past, which, although it is fixed as far as its physical characteristics are concerned, is not absolutely fixed in so far as other eternal objects can be applied to it through conceptual reinterpretation.

In neither the prehensions nor the prehendibility of actual occasions is any ontologically fundamental discreteness involved. Whitehead's emphasis on atomicity suggests such discreteness, however. In so far as one follows the suggestion, he runs into paradoxes and difficulties of interpretation. Not only is a historic route impossible if actual entities are in any sense discrete, but enduring objects are also impossible. Whitehead lays the foundation for the solution of most of these difficulties in his emphasis on continuity, but he did not carry out the elaboration of the doctrine with sufficient clarity. He takes continuity for granted, and in many passages seems to be satisfied with Aristotle's insufficient analysis of continuity as being merely infinite divisibility. This will not do. Infinite divisibility yields only a dense series, not a linear continuum. All the units in a dense series are discrete, for every dense series is denumerable. A dense series yields only pseudo-continuity. Whitehead's method of extensive abstraction starts from a continuum of the linear variety, but whenever he treats of continuity in *Process and Reality*, he is satisfied with the Aristotelian variety, which is the pseudo-continuity of the dense series.

Whitehead did not sufficiently analyze the problem of the relation between the continuous and the discrete. To say that potentiality is continuous and actuality is discrete does not solve the problem. Yet the emphasis on the atomicity of actual entities at least suggests that this is the only preferred solution. If there is *any* ultimate discontinuity in a cosmology, then the discrete not the continuous is ontologically fundamental. Continuity cannot be obtained by a summation of discrete parts even if there are only two such parts. Discrete parts yield only a pseudo-continuity at best. You cannot have it both ways, but Whitehead seems to insist on having it both ways. His emphasis on the atomicity of actual entities gives them discrete characteristics, but they are continuous too, as they are extensive. This strikes the present writer as a highly sophisticated way of eating one's cake and having it too.

Such a conclusion is not necessary within the framework of Whitehead's philosophy if due emphasis on the perception of causal efficacy is retained throughout. The perception of causal efficacy is the experience of a full continuity, and this continuity lies at the basis of Whitehead's empiricism. Without it, the cosmology will not work. With it the cosmology will work, and the basic empirical evidence for continuity is the perception of causal efficacy. If sufficient emphasis is given to the empiricism of the doctrine and to the continuity inherent in this empiricism, then the atomicity of actual entities disappears, and with it, many of the paradoxes of Whitehead's cosmology disappear.

WHITEHEAD'S "ACTUAL OCCASION"

STEPHEN C. PEPPER*

In Whitehead's philosophy, as this was developed in *Process and Reality* and in his later works, the "actual occasion" is presented as the pivotal actuality in the world and in human experience.

This term is one of his own coinage. An immediately discovered difficulty in approaching Whitehead is the large number of words he coined to explain his system. Some of these we will encounter soon. It is probably just as well to give a preliminary idea of what he means by them right off. "Actual occasion" is suggestively descriptive of what he is referring to. The reference is to an instance of what we experience as going on now. It is very much what we commonly mean by an event going on at the time we are experiencing it. The experiencing of it Whitehead calls "feeling" and means the full qualitative immediacy of the experience – just what we feel emotionally, sensationally or otherwise as it is going on. But "feeling" for Whitehead does not necessarily involve consciousness. Conscious feeling is a special mode of feeling.

When an actual occasion is past it continues to have a sort of actuality. But its actuality is no longer that of a living energizing process. It becomes a stubborn historical fact which never changes again and which is consequently often referred to by Whitehead as "immortal." All the facts of perception and of scientific observation as well as of history are these stubborn immortal facts of the past. This is not so paradoxical as it first sounds when we realize that it takes time for an environmental object under observation to stimulate the sense organ of an observer so that inevitably any

* Philosophy Department, University of California, Berkeley. Visiting Professor of Philosophy, Tulane University, 1961.

observed fact is somewhat in the past at the moment of observation. And Whitehead takes the evidences for energy transmission in the physical world seriously.

There is one other actual thing in the universe, according to Whitehead, and that is God, who shares some of the characteristics of the actuality of an occasion going on now and some of those of the immortal past. Our concern in this paper will not involve Whitehead's concept of God, so I will not expand any further on this.

In all there are three types of actual entities – actual occasions going on now, occasions of the past or stubborn facts, and God. Sometimes Whitehead uses "actual occasion" and "actual entity" as synonymous. But in general practice he means by "actual occasion" only the felt immediate event. This is what I shall be referring to as the topic of this paper.

Two other important concepts of Whitehead's system must be referred to at this time – namely, "eternal objects" and "extension." By eternal objects he means what in the philosophical tradition are called Platonic forms or universals. His forms do have certain distinctive characteristics, and that is why he gives them a separate name. He does not want them confused with any of the traditional treatments of them. However, these distinctive characteristics will not concern us in this paper. By "extension" he means space-time. This is what ordinary common sense would also mean by it. But Whitehead differs from at least one ordinary conception of space-time in that he does not regard it as "actual." Neither eternal objects nor space-time are actual. But nevertheless for him they are both "real." So, "actuality" and "reality" must be distinguished. It would not be entirely unfair to say that actual entities have a first-class status of reality whereas eternal objects and extension and other merely real things have a second-class status of reality. For instance, Whitehead sometimes refers to these as "abstractions." They would, of course, be abstractions from actual entities.

With the foregoing preliminaries, let me now return to my initial statement that for Whitehead's mature philosophy, the actual occasion is the pivotal actuality in the world and in human experience. This fact has come to have more and more weight for me in interpreting the significance of White-head's philosophy.

In his earlier works he was occupied with the analysis of natural science and its methods. He was chiefly concerned in describing objects and events and how we come to know them. The dependence of scientific concepts on qualitative obser-vation, however, absorbed his attention more and more deeply. He found a way of developing the primitive concepts of geometry – point, line, and plane – by "extensive ab-straction" from immediate qualitative experience. He became critical of the sense-data theory as a basis for connecting scientific concepts with sensory observations, and critical of the identification of scientific concepts, such as mass, centi-meter, and second, with cosmic actuality. His criticisms be-came crystallized into pungent phrases for typical fallacies – the fallacy of 'simple location' and that of 'misplaced con-creteness.' He gravitated more and more towards a recog-nition of the richness and the primacy of qualitative feeling as that which actually occurs and goes on in the universe. The expression of this insight we have in its full bloom in *Process and Reality,* culminating in the fertile concept of a qualitative concrescent actual occasion.

I have come to think that perhaps the actual occasion might be regarded as the root metaphor of his mature philo-sophy – the germ out of which his philosophical categories mainly sprout. If so, this should probably be regarded as a new root metaphor, different from those which generated the great traditional schools of formism, mechanism, organicism, and contextualism. And if so, the actual occasion has more than a technical interest. It could be the source of a new philosophical school – a new world hypothesis.

Until recently I had thought of Whitehead as an eclectic – one who combines features of different world theories based on different root metaphors to the confusion and incoherence of the resultant hodge-podge. I conceded that he had done an astonishingly clever job of concealing the joints where the borrowed fragments came together, and giving an impression of total coherence. But his was nevertheless, I held, only an eclectic theory, inconsistent and basically confused in conception. I still find eclecticism in his theory. But one's attitude changes radically if one comes to believe the author is struggling with something novel. His eclecticism then becomes of secondary importance. The thing of first importance is this new root metaphor. What is it? How fruitful is it likely to be? What are its proper categories? How consistently do they spread their interpretations over the world, and how adequately are they performing?

In this paper, I shall restrict myself to the first question – What is this root metaphor? – and more specifically to Whitehead's description of the phases and features of the actual occasion. At the end I shall endeavor to make some criticisms of a constructive nature.

I believe the root metaphor is what we ordinarily call a purposive act. Such an act, as we watch it going on, is through-and-through qualitative. It has a distinctive unity, drive, and aim. It has a degree of extensiveness both spatially and temporally. It is closely engaged with an environment with which it is in constant contact. It has simpler and more complex forms. It seems to involve in it every qualitative feature we can imagine. If one has come to believe that the actuality of the universe is qualitative throughout, and any appearances to the contrary are abstractions from this qualitative base, a purposive act is as fair a sample of such a whole as we are likely to find.

I am not saying that Whitehead consciously selected the purposive act as the model for his description of the actual occasion. He came at the realization of the basic qualitative nature of actual things by another route. He did not start with feelings and emotions and urges for satisfaction as indubitable elements of actuality. He started with mathematical and scientific concepts and found these meaningless unless they reached their meaning in the qualitative immediacy of feelings and satisfactions. Then he found these feelings embedded in a structural process which he delineated with some care. The result came out as something very close to a detailed behavioristic description of a purposive act introspectively interpreted. He called it the "actual occasion."

How then does Whitehead describe an actual occasion? It

is a process with a duration and feeling all its own. It may be mine or yours or elsewhere in the universe. But as mine, it may be illuminated for me with consciousness, though not all actual occasions are conscious ones. In fact, only a few. But all are qualitative with feelings like conscious ones, which latter differ only in having a special complexity. In the flow of our actual experience, we are sometimes conscious and sometimes not conscious of what we are doing. Both are equally filled with feeling qualities. But presumably it is only from our periods of conscious feeling that we directly describe the features of an actual occasion. This is where we get our direct evidence for its nature, though there is plenty of indirect evidence from observing the effects of other actual occasions which can amplify our description of a given one as it goes on.

With this much in mind, let me now give a list of the main features of an actual occasion. Then I can take these features up one by one.

1. It is a *process* qualitatively felt over a duration and going through successive phases.
2. It is an integrative process, and hence called by Whitehead a *concrescence*.
3. It culminates in a final phase of complete integration, called *satisfaction*.
4. It is directed towards this satisfaction by a *subjective aim* which acts as a final cause.
5. It is initiated by an efficient cause which Whitehead calls *physical prehension*.
6. It contains an element of novelty which comes from *conceptual prehension*.
7. It is a *total unit* while in process and indivisible in terms of physical space-time. It is, or is like, James' 'specious present.'
8. When satisfaction is achieved, it ceases to be in process and becomes a fixed and stubborn fact in the past. It thus attains an immortal reality like all historical facts. But also it becomes divisible in terms of physical space-time, perceivable by physical prehension, a source of data for actual occasions in process, and the source of efficient causation.

This last feature of the factual immortality which accrues to an actual occasion when it has achieved its satisfaction, is something I shall not get involved with in this paper, except as it may bear on the features of an actual occasion in process. Only the latter is fully actual. The immortal past is real only in a sort of secondary sense, and perhaps only because of a peculiar tenet of Whitehead's by which the whole of the world's past is embodied positively or negatively in each actual occasion's process of prehension. This tenet may be a sign of Bergson's considerable influence upon him. Anyhow, we are justified in disregarding it as a feature of an actual occasion in process.

Put the other seven main features of an actual occasion together, and they afford a fairly detailed description of a goal seeking purpose in action. There is a drive (the efficient cause) which sets in motion a process of integrating means towards an end aimed at and which culminates in a feeling of satisfaction on reaching the goal. And then the action terminates. Moreover, as an actual purpose it is qualitative throughout (not a behavioristic or physiological conceptual scheme), and the feeling of the passage of time is a qualitative duration (not a physical measurement in minutes and seconds). With such an interpretation, Whitehead's actual occasion is brought into contact with our everyday living. We can see what he is talking about and check up his descriptions against our experiences of the things he describes.

I shall now proceed to examine in detail the first seven main features listed above to see how they operate together, and ultimately to see whether we may conclude that Whitehead did have a purposive structure in mind, and, if he did, how accurately he described it, and how far it may be regarded as a new root metaphor for a new theory.

(1) We begin, then, with Whitehead's conception of a process. This is a felt duration genetically divisible (PR 433) into successive phases. These phases are qualitative throughout, permeated with emotions, sensations, and the like. So much is clear. But when we then ask how long is the duration of an actual occasion, we are blocked. One of the reasons for frustration here is the meagreness of concrete illustration in

Whitehead's writings. We should like to ask whether the duration of the process of an actual occasion would be the equivalent in real life of a scene in one of Shakespear's plays – say, the scene between Hamlet and his mother when Polonius was killed. Or must it be shorter like the buying of a postage stamp at a Post Office? Or even shorter than that?

That it must be fairly long one would surmise from Whitehead's numerous descriptions of what goes on in the integrative phases of concrescence. Near the beginning of Chapter 14 of *Adventures of Ideas* he writes, "The intermediate phase of self-formation is a ferment of qualitative valuation. These qualitative feelings are either derived directly from qualities illustrated in the primary phase, or are indirectly derived from their relevance to them. These conceptual feelings pass in novel relations to each other, felt with novel emphasis of subjective form. The ferment of valuation is integrated with physical prehensions of the physical pole. Thus the initial objective content is still there. But it is overlaid by, and intermixed with, the novel hybrid prehensions derived from integration with the conceptual ferment. In the higher types of actual occasions, propositional feelings are now dominant. The enlarged objective content obtains a coordination adapting it to the enjoyments and purposes fulfilling the subjective aim of the new occasion."

We can recognize in this technical Whiteheadian description experiences of our own by which we reach a solution for a complex situation after a first perception of its physical impact. A child comes in sobbing with bleeding knees after a bad fall. There are some things immediately to do. Give the child some comfort and some confidence. Draw some warm water and wash out the dirt. Then the ferment of more or less relevant thoughts. Associations springing out of these. How deep is the abrasion? When will it stop bleeding? What to put on it? Where are there some bandages? Would a handkerchief stay on? Where is the alcohol or bay rum? Will it smart? Is the child hurt elsewhere? Does this call for a doctor? And then in reference to the look of the wound and the state of the child, irrelevant thoughts get eliminated, and relevant ones begin to get in order, and a purposive aim emerges which the situation itself requires, and in awhile it all comes to a satisfactory ending as far as this particular episode is concerned.

This sort of occurrence checks well with the Whiteheadean description quoted. But it will take some bit of time. Its duration by a clock will be at least fifteen or twenty minutes. Of course, it was not lived through by the clock. It was experienced as a qualitative duration with tensions and relaxions of feeling. But looked back upon, and checked up by the clock, it must have taken fifteen or twenty minutes. A duration qualitatively experienced is not physical clock time, but it can be correlated relevantly with the movements of the hands of a clock. And such an occurrence would have a duration of a fair length of time.

Another description by Whitehead of an occasion, one of the few vivid concrete ones, is of an angry man. The continuity of his anger is described as passing through a number of occasions. The problem of how this quality passes from occasion to occasion in a qualitative continuity is Whitehead's problem in this instance. The reference is to Chapter 11, section 14 of *Adventures of Ideas*. "Suppose at some period of time" writes Whitehead, "some circumstance of his life has aroused anger in a man. How does he now know that a quarter of a second ago he was angry? Of course, he remembers it; we all know that. But I am enquiring about this very curious fact of memory, and have chosen an overwhelmingly vivid instance. The mere word 'memory' explains nothing. The first phase in the immediacy of a new occasion is that of conformation of feelings. The feeling as enjoyed in the past occasion is present in the new occasion as datum felt. . . Thus if A be the past occasion, D the datum felt by A with subjective form describable as A angry, then this feeling – namely, A feeling D with subjective form of anger – is initially felt by the new occasion B with the same subjective form of anger. The anger is continuous throughout the successive occasions of experience. This continuity of subjective form is the initial sympathy of B for A. It is the primary ground for the continuity of nature. Let us elaborate the consideration of the angry man. His anger is the subjective form of his feeling for some datum D. A quarter of a second later he is, consciously, or unconsciously, embodying his past as a datum in the present, and maintaining in the present the anger which is a datum from the past. Insofar as that feeling has fallen within the illumination of consciousness, he enjoys a non-sensuous perception of the past emotion. He enjoys this emotion, both objectively, as belonging to the past, and also formally as belonging to the present. This continuation is the continuity of nature. I have labored this point, because traditional doctrines involve its denial. Thus non-sensuous perception is one aspect of the continuity of nature."

There is a lot in this passage that is outside our immediate question as to the duration of a single occasion. I quote it all so that we may have before us the whole of the reference, and also so that I can refer to it later for other things. But its present relevancy is to emphasize how very short a duration for Whitehead may be. It is timed by him here as $\frac{1}{4}$ second. He pictures for us here a man who has become angry and stays angry quite awhile. Let us imagine he stays angry for five minutes. Then on this analysis of Whitehead's this quality of anger has passed through $5 \times 60 \times 4 = 1200$ occasions $\frac{1}{4}$ second long. On this mode of analysis, the anxiety of the mother described above tending to her sobbing child would likewise have passed through many hundreds of occasions. Can Whitehead really mean this?

Clearly there is a discrepancy between the description of a concrescing occasion with phases of ferment and sifting and integration towards a final satisfaction, and this description of a mere continuity of a quality through time.

Could the discrepancy be harmonized by recognizing occasions of quite different durations – very short ones and long ones? Whitehead sometimes seems to sanction this solution. But then how do we know when one occasion stops and another begins? Are there possibly some twelve or twenty thousand short occasions of anxiety contained in the one occasion of the mother's tending her child? Can lesser occasions be included in larger ones? There are passages that would seem to support this solution, too.

But there is one set of considerations that would seem decisive for a fixed cosmic span for the duration of a single occasion. This emerges as a consequence of Whitehead's interpretation of the bearing of the special theory of relativity on his philosophy of actual occasions. On this theory, whatever has a causal influence upon an occasion is in the past of that occasion. According to the special theory of relativity, the highest speed in the world by which causal impetus can be transmitted is that of the velocity of light. It follows that contemporary actual occasions are all those which cannot have a causal influence on one another because to have such influence the impetus would have to be transmitted faster than the velocity of light. The present in time, therefore, has a certain thickness which is its cosmic duration. The past occasions are those that can have a causal influence on present ones. The future occasions are those which present occasions can causally influence later. Cosmic duration at any present times includes all those contemporary occasions which can have no influence on each other. It follows from these considerations that the thickness of cosmic duration for a given actual occasion cannot be very great. Whitehead may mean strictly what he says when he estimates the angry man's actual present occasion as $\frac{1}{4}$ second.

This interpretation seems to be borne out in numerous other passages. For instance, "Memory is a perception relating to the data from some historic route of ultimate percipient subjects M_1, M_2, M_3, etc., leading up to M which is the memorizing percipient" (PR 184). This description parallels exactly the 'angry man' example. This is clearly a succession of actual occasions, presumably about $\frac{1}{4}$ second long each. And then (PR 197) he describes the utterance of a sentence: "We are carried on by our immediate past of personal experience; we finish a sentence *because* we have begun it. The sentence may embody a new thought never phrased before, or an old one rephrased with verbal novelty. There need be no well-worn association between the sounds of the earlier and the later words. But it remains remorselessly true, that we finish a sentence *because* we have begun it. We are governed by stubborn fact."

Since "stubborn fact" for Whitehead is always in the causal past outside the concrescent duration of an actual occasion,

Whitehead is obviously committed here to the view that the earlier part of a sentence during utterance is outside the actual occasion in process. The duration of the occasion is limited to the tail end of the sentence, and must be very short. The earlier part is all in the past and perceived in memory as stubborn fact.

In all such passages the duration of an actual occasion is short. And it is definite that a short actual occasion cannot be included within some longer actual occasion, because what is past is forever past and cannot be reenlivened into a present feeling without violating one of Whitehead's basic principles – that of the atomicity and indivisible immediacy of a concrescing occasion.

On the other hand, when Whitehead is describing the process of concrescence, the duration of an actual occasion spreads out broadly with many stages. In this mood he can say, "An actual occasion is nothing but the unity to be ascribed to a particular instance of concrescence... [with] three stages in the process of feeling: (i) the responsive stage, (ii) the supplemental stage, and (iii) the satisfaction. The satisfaction is merely the culmination marking the evaporation of all indetermination... But the process itself lies in the two former phases. The first phase is the phase of pure reception... The feelings are felt as belonging to the external centres, and are not absorbed into private immediacy. The second stage is governed by the private ideal, gradually shaped in the process itself; whereby the many feelings, derivatively felt as alien, are transformed into a unity of aesthetic appreciation immediately felt as private. This is the incoming of 'appetition' which in its higher exemplifications we term 'vision.'" And there are two subordinate phases in this second phase – the "aesthetic supplement" in which "private immediacy has welded the data into a new fact of blind feeling" and an "intellectual supplement" which may introduce a conscious hypothesis of the aim "eliciting, into feeling, the full contrast between mere... potentiality and realized fact." And he adds, "The actualities are constituted by their real genetic phases. The present is the immediacy of the teleological process" (PR 323–7).

Now all this cannot go on in $\frac{1}{4}$ second. Here he is describing a full purposive act. And, let me repeat again, he cannot escape the discrepancy between these accounts by considering the $\frac{1}{4}$ second actual occasions as elements within the long term occasion. For each actual occasion is an atomic whole coordinately indivisible into its phases.

So right at the beginning we meet with a serious difficulty. If a purposive act in its full integrity from its initial aim to its terminal satisfaction is intended as the concrete observable model for the description of an actual occasion, then the duration of the occasion must often be quite long. On the other hand, if an actual occasion is the juncture between past and future, such that other actual occasions contemporary with it cannot have a causal influence upon it, then it must be very short. Moreover, if it must be limited by the duration of a specious present, then again it must be rather short. However, let us overlook this discrepancy for the present and continue our examination of the actual occasion as he describes it in its successive stages.

(2) He is emphatic on the integrative concrescence of the actual occasion. From its initial phases it moves steadily towards a satisfaction which brings together all the relevant qualities leading to the goal and eliminates the irrelevant ones. There are degrees of relevancy for the qualities that enter the occasion to be integrated in the concrescent process. These degrees of relevance constitute "valuation" for Whitehead. This is what we should expect in purposive activity. Elements become valued in proportion to their contribution to the goal of the purpose. If they do not positively contribute, they are eliminated. The process of elimination is termed by Whitehead 'negative prehension.' He points out that negative prehensions make a distinct difference to the concrescent process. To avoid something is just as important as making use of something in the course of reaching purposive satisfaction. But what is avoided gets negatively valued in the process. Moreover, what is accepted may be of more or less importance in the attainment of the goal and so becomes variously valued. Finally, the total successful act is an

integrated one in which every item accepted has its proper function which it performs.

(3) The result terminating the act is a satisfaction, the third feature in our list describing an actual occasion. This is the feeling of the attainment of the goal through the integrated functioning of all the contributing elements.

Whitehead seems to conceive this as gathering up all the contributing elements into a unitary whole felt all at once as a whole. Thus satisfaction is not only the terminal phase of a concrescence, but the embracing of total constructive achievement in the feeling of the integrated whole. The concept of the specious present helps to understand what this may mean. There is past, present and future, a span of duration, in the grasping of a phrase of music or of a sentence. The successive phrases are there, but felt also as all there at once. But for Whitehead the irrelevances get all eliminated and the relevant functions get all integrated in the final satisfaction.

If this is an actually experienced purpose being described, it is clearly a positive appetitive purpose of the aesthetic or intellectual type – perhaps more precisely of the aesthetic type. Chapters 17 and 18 of *Adventures of Ideas* are illuminating in this respect. In Sections I and II, Chapter 17, Whitehead says, "Thus the parts contribute to the massive feeling of the whole, and the whole contributes to the intensity of feeling of the parts. . . In other words, the perfection of Beauty is defined as being the perfection of Harmony, and the perfection of Harmony is defined in terms of the perfection of Subjective Form in detail and in final synthesis. . . In order to understand this definition of Beauty, it is necessary to keep in mind three doctrines which belong to the metaphysical system *in terms of which the World is being interpreted in these chapters*" (my italics). He then proceeds to explain these three doctrines. The first refers to the qualitative immediately felt contents of an actual occasion. The second, to "the unity of the immediate occasion in process of formation." The third, to "the final autonomy of the process." These are a selection from the features of the actual occasion we have already enumerated. He is very nearly asserting here that the actual occasion is his root metaphor, and that the essential features of an actual occasion constitute his basic categories of interpretation. But our attention for the moment is upon the second of his three doctrines, the unity of the immediate occasion, and its reference back to the "perfection of Harmony... in final synthesis" and the equating of this by definition with the "perfection of Beauty."

Whitehead's description of the purposive action of an actual occasion corresponds closely with the purposive activity of a creative artist. A painter or a poet receives his impetus from some environmental stimulus or some experience out of his past. Presently his ideas begin to take form and a more and more clearly envisaged aim emerges from his creative activity. For this picture or this poem some suggestions are relevant, some not. The latter are eliminated, the others are brought to function harmoniously together. Finally there is the synthesis of all these details in the harmony of an integrated whole which somehow incorporates the whole process that was gone through in the finished picture or poem. This integrated whole is felt as that unique satisfaction of that fulfilment. And then the picture or poem drops into history, as a stubborn fact that may be referred to later, or may help to inspire some other creative act in its future. For such a purposive act Whitehead's descriptions of actual occasions on a broad scale fit excellently. (But it must be interpolated that all purposive acts are not of this supreme type.)

(4) This brings us to the subjective aim. A goal-seeking purpose acquires an aim from the start. The aim is only sometimes an intellectually clear concept. In primitive purposive activity it arises from the impetus of the drive which demands a certain terminus for its satisfaction. But Whitehead's actual occasion as broadly presented is far from a primitive purposive act. It gathers its impetus from diverse sources in its past and also from a spontaneity inherent in itself. The aim is determined by the convergence of these diverse impulses towards an integration of their contents or "feelings" as he calls them. There are intimations in Whitehead's references to the subjective aim that this is not fully

exhibited till it is achieved in the final satisfaction. The aim is inherent in the concrescent process from the beginning. It is a real potentiality of every earlier phase of the purposive process towards the determinate goal of the integration of all the relevant contents of an actual occasion.

This feature of a real potentiality imbedded in the purposive process is something to be stressed. All goal-seeking purposes have it. It passes unnoticed in most accounts of such activity because of the strong mechanistic bias of scientific tradition which rejects any hint of potentiality as committing one to something Aristotelian. It is one of Whitehead's bold insights that a real potentiality is inherent in a purposive process. And since for Whitehead all the processes of nature down to the simplest interactions of subatomic elements are actual occasions, it follows that real potentialities extend throughout nature. It is difficult otherwise to explain the grounds of prediction.

A subjective aim institutes a real potentiality in an actual occasion. Whitehead brazenly names it the "final cause," an Aristotelian term, so that there will be no question that he means what he says. The subjective aim directs the concrescent process to the integrated satisfaction. The terminal phase is incipient in the initial phase and bound to be achieved. It is a real inherent potentiality from the beginning. It is not a pure possibility like that of something being green or gold or conscious sometime, or possibly not. Nor is it simply a mental hypothesis, what he calls a "proposition." A real potentiality is a feature of a natural process connecting and directing an earlier phase of the process to a later and a terminal phase. The subjective aim is a final cause embedded in an actual occasion.

(5) But a final cause is only part of the dynamics of an actual occasion. The other part is what he calls the efficient cause, borrowing another Aristotelian term. This is the causal impetus that is transmitted from a past occasion to a present one. He amalgamates the efficient cause with the process of perception. When an actual occasion has a perception of a quality, it is experiencing an efficient cause operating upon the actual occasion by an environing occasion. This latter

occasion is necessarily in the past of the occasion in process, since it takes some time (if only a very little at the velocity of light) for the impulse to be transmitted. Of course, any complex actual occasion is subject to efficient causation from many environmental sources – all that are relevant to (that is, "positively prehended" by) the actual occasion in process.

(6) It might be thought that the convergence of these causal impulses upon the occasion in process would be sufficient to give it the dynamics, and even the final cause for its con-crescence. But Whitehead is impressed with the uniqueness and the novelty of a purposive act. Again this indicates to me his taking as his ultimate model a creative purposive act such as that of an artist, inventor or imaginative scientist. He is so impressed that he will not leave the novelty to be explained by the uniqueness of a configuration of efficient causes con-verging on a complex purposive act. He introduces what he calls a "conceptual prehension" which is the intuition of an original idea never actualized in the world before and so in-capable of acting as one of the efficient causes out of the natural environment. Since, however, for Whitehead an actual entity must always be found as a "reason" for any occurrence in the world, this original idea must have some actual entity as its source. This he finds in God whom he describes as an actual entity with a primordial status which conceptually prehends all pure possibilities. So all original ideas are in this sense attributable to God, and yet White-head is careful to make out that God does not himself put an original idea into an actual occasion. The spontaneity of the occasion is required for this. It may be said that God in his primordial status does not know when an original idea will become actualized until the favorable occasion occurs which takes it in.

However, we are not to be concerned here with White-head's conception of God except as it bears on the source of the original idea which is a feature of uniqueness in every actual occasion. This conceptual prehension, entirely free of efficient causation, is incorporated into the final cause or subjective aim of the purpose. Thus every purposive act in its concrescence is a unique creative act in the fullest sense of the term. And the integrative satisfaction attained is a unique creative achievement.

(7) It just remains to be added that in Whitehead's conception the total act of concrescence is a physically indivisible unit. It is an overarchingly given duration felt all at once. It is not, except in retrospect, when it is already past and dead, an extent of clock time divisible into minutes and seconds. This feature has already been touched upon. It is a reference to the experience of temporal spread in what is called the specious present. An actual occasion is apparently conceived by Whitehead as embraced in a specious present. Moreover, the felt quality of it is finally identified with its integrated terminal satisfaction. The preparatory phases of concrescence become absorbed, and transcended in the total integrative whole.

One of Whitehead's motives in insisting upon the indivisible unity of an actual occasion is to provide for atomicity in nature. An adequate world theory must, he maintains, provide for both the continuity and the atomicity which are involved in the physical treatment of nature. Efficient causality together with his theory of extension, Whitehead believes, will take care of his continuity of nature. But the physically indivisible unity of an actual occasion is needed to take care of atomicity.

Our analysis of the actual occasion must stop here. A few critical comments remain to be made.

First, there seems to be a double description of the actual occasion running through *Process and Reality*. There is a narrow and a broad description. The one seems required by his theory of space-time and some of its consequences. The other is required to fit immediate experience as qualitatively lived through. Only the second interpretation has an analogy to purposive activity. His considerable, not to say predominant, emphasis on this broader description makes one wonder if Whitehead is not here breaking out a new root metaphor. If so, we may be witnessing here the emergence of a new world hypothesis, a new school of philosophy. This paper has been an exploration of this possibility.

Second, it must be conceded that Whitehead does not

follow this root metaphor of the purposive act in process very consistently. He is still very much of an eclectic and even sometimes practically says he intends to be. At other times, as in one passage earlier quoted, he seems to be saying that the actual occasion is his root metaphor in terms of which he seeks to interpret the world.

Third, accepting Whitehead's broad description as an incipient attempt to take the purposive act as a root metaphor, it is clearly an inadequate and arbitrarily selective description of such an act. It is limited to the creative purposive act of a man of creative imagination. This is a highly complicated and an exceptional type of purposive activity. There are many other forms of purposive activity. Moreover, the description is limited to goal-seeking purposive activity and is neglectful of aversive activity (except under the thin heading of negative prehension) which is purposive activity without a goal but in avoidance of an object of apprehension.

Other lesser shortcomings might be mentioned. But these are sufficient to show the direction such criticisms would take. The direction is to show what modifications in Whitehead's treatment of the actual occasion might lead to a new world theory more adequate than his, if his is interpreted as an attempt to use the purposive act as a root metaphor. What he has succeeded in doing with a very limited and defective use of this root metaphor is perhaps a token of what might be done with it in a more extended way. And perhaps part of the astonishing popularity of his difficult books including, amazingly, *Process and Reality* itself, which has just come out in a paper-back edition, is a sense that here is a new approach to the interpretation of the world, that the author is a genuinely creative philosopher with a glimpse of an important insight.

THE PHILOSOPHY OF CHARLES
HARTSHORNE

ANDREW J. RECK

"Whitehead has, I have no doubt," asserts Charles Hartshorne, "achieved the major metaphysical synthesis of our
day." [1] If the ideas of a philosopher reveal their meaning and
value through their incorporation and modification in the
thinking of later philosophers, then the significance of Whitehead's philosophy may already be discerned in the adventures
of its basic ideas in the works of contemporary thinkers who
have come to comprise what my colleague, Professor Robert
C. Whittemore, has aptly called "The School of Whitehead."
Perhaps no contemporary thinker matches Charles Hartshorne in the adoption, adaptation and elaboration of Whiteheadian ideas, though, let us add, he owes as much to C. S.
Peirce as to Whitehead. [2] Neither limiting the source of his
ideas to Whitehead, nor refraining from novel variations on
his own part, Hartshorne has nonetheless insisted that there
is a present convergence of views in metaphysics, constituting
in effect a consistent yet viable movement, and that this
movement is most adequately and coherently articulated in
the philosophy of Whitehead. "I take it," Hartshorne writes,
"that Bergson, James, Fechner, Alexander, Whitehead,

[1] Charles Hartshorne, "The Compound Individual," *Philosophical Essays for
Alfred North Whitehead* (London, New York, Toronto: Longmans, Green and Co.,
1936), p. 212. In subsequent citations this essay will be represented by the abbreviation "CI."

[2] It should be recalled that Hartshorne was one of the editors of Peirce's papers.
See *The Collected Papers of Charles Sanders Peirce*, edited by Charles Hartshorne
and Paul Weiss, Vols. I–VI (Cambridge: Harvard University Press, 1931–1936).
Hartshorne is not alone in perceiving affinities between Peirce's philosophy and
Whitehead's, even though Whitehead never studied Peirce. Some of the commen
tators on Peirce concur with Hartshorne's viewpoint. For a provocative discussion of "astounding" similarities between the philosophies of Whitehead and
Peirce, see James Feibleman, *An Introduction to Peirce's Philosophy Interpreted
as a System* (New York and London: Harper and Brothers Publishers, 1946),
pp. 459–463.

Varisco, Scheler, Ward, Boutroux, Montague, Parker, Garnett, Hocking, Boodin, and others, including the present writer, are in a certain rough agreement that is somewhat more striking and representative of metaphysics since about 1850 than is any other trend." [1] Of this new trend, epitomized in the title of one of his books, *Reality as Social Process*, Hartshorne remarks: "Whitehead is its Einstein, Leibniz was its Newton. . ." (RSP, 31).

Hartshorne's philosophy may be viewed as a creative variation of the new trend in metaphysics. His first book, *The Philosophy and Psychology of Sensation*, introduces into psychological theory principles of continuity and aesthetic feeling espoused by the new metaphysics in the philosophy of mathematics and of nature, and the upshot is the novel doctrine of the affective continuum. [2] Hartshorne's next book, *Beyond Humanism*, announces the movement as "a genuine integration of all the modern motifs" culminating in a new theology, which he designates "theistic naturalism or naturalistic theism," and which he presents in contrast with and in opposition to its great contemporary rival, "non-theistic" humanism. [3] *Man's Vision of God* undertakes to formulate the logic of the new theism and to demonstrate its superiority to the classical synthesis of Thomas Aquinas. [4] In these works Hartshorne expounds and defends a metaphysics of panpsychism and he also suggests what he subsequently develops more fully as a theology of panentheism. In his Terry Lectures at Yale, later expanded and published as *The Divine Relativity*, [5] Hartshorne systematically formulates the panen-

[1] Charles Hartshorne, *Reality as Social Process; Studies in Metaphysics and Religion* (Glencoe and Boston: The Free Press and Beacon Press, 1953), p. 131. In subsequent citations this book will be represented by the abbreviation "RSP."

[2] Charles Hartshorne, *The Philosophy and Psychology of Sensation* (Chicago: The University of Chicago Press, 1934). In subsequent citations this book will be represented by the abbreviation "PPS."

[3] Charles Hartshorne, *Beyond Humanism; Essays in the New Philosophy of Nature* (Chicago and New York: Willett, Clark and Co., 1937), pp. viii–x. In subsequent citations this book will be represented by the abbreviation "BH."

[4] Charles Hartshorne, *Man's Vision of God and the Logic of Theism* (New York: Harper and Brothers Publishers, 1941). In subsequent citations this book will be represented by the abbreviation "MVG."

[5] Charles Hartshorne, *The Divine Relativity: A Social Conception of God* (New Haven: Yale University Press, 1948). In subsequent citations this book will be represented by the abbreviation "DR."

theistic conception of deity, a conception which owes much to Whitehead.[1] Then, in *Philosophers Speak of God*, Charles Hartshorne, in joint authorship with William Reese, presents and discusses all the important possible conceptions of deity, arguing in the comments and in the opening and concluding chapters for the validity of panentheism.[2]

In this paper I propose (1) to explore the three main aspects of Hartshorne's thought: (a) the affective continuum, (b) panpsychism and social realism, and (c) panentheism; and (2) to suggest, without detailed demonstration, how these ideas are specifications and variations of themes characteristic of the new trend in philosophy of which, admittedly, Whitehead is the leading exponent. Above all, I do not intend to insinuate that Hartshorne's thought is merely derivative. As my discussion of his ideas should make plain, Hartshorne is a creative metaphysician of extraordinary ability, remarkable for his logical rigor as well as his speculative originality. When the ideas of a philosopher are prehended by a second philosopher of equal genius and with creative intent, they undergo a process of assimilation which, subordinating them to the subjective form of the second philosopher, imbues them with novelty. Further, as novelty in part is a function of contrast, the adventures in which ideas engage disclose, in the new contexts, their hitherto non-existent value and significance. So it is with the ideas we shall now explore.

I. THE AFFECTIVE CONTINUUM

In *The Philosophy and Psychology of Sensation* Hartshorne investigates the concept of sensation in the light of ideas which, despite fruitful employment in the interpretation of the physical sciences, had not yet found application to psychological science, where unfortunately obsolete concepts, incom-

---

[1] See Charles Hartshorne, "Whitehead's Idea of God," *The Philosophy of Alfred North Whitehead*, edited by Paul Arthur Schilpp (New York: Tudor Publishing Company, 1951), 2nd edition, pp. 515–559.

[2] Charles Hartshorne and William L. Reese, *Philosophers Speak of God* (Chicago The University of Chicago Press, 1953). In subsequent citations this book will be represented by the abbreviation "PSG."

patible with the philosophic principles required by the recent revolutions in the physical and biological sciences, still held sway, particularly as regards sensation. Hartshorne's theme, therefore, is that ". . . the application of scientific and rational principles to the sensory qualities results in a new theory of these 'immediate data of consciousness,' considered both in themselves and in relation to their physical stimuli, organic conditions, biological significance, and evolutionary origin" (PPS, 6).

To this new theory of the immediate data of consciousness, the prevailing alternative, an alternative continued in psychology in spite of its obsolescence in physics, is materialism. In the modified sense Hartshorne accepts, materialism supposes that ". . . matter, in what may be called the 'bad' sense, exists" (PPS, 11). Materialism posits the existence of atoms, discontinuous, discrete, independent bits of matter, devoid of feeling and life, isolated except for accidental external relations, timeless and unchanging with respect to internal constitution and hence without growth or evolution. This conception of matter, fostered by Newtonian physics, has been rejected by the recent revolutions in physics, which, according to Hartshorne, give ". . . more encouragement in detail to the Whiteheadian conception of events composed of aesthetic feeling as the materials of all nature than Newtonian physics could give to Leibnizian monadology" (PPS, 16–17). But if materialism is repudiated in physics, it must also be repudiated in psychology, where materialistic modes of thought still linger, not only, as is obvious, in the rise of behaviorism, but also in the theories of sensation. For so long as psychologists adopt theories of sensation which treat sensations as discrete data, separated from each other, sharply distinguished from feeling, and isolated on the one hand from the internal neural processes of the sentient organism and on the other hand from the external physical stimuli, they are, unwittingly perhaps, adhering to an outmoded materialism.

But what theory does the application of the new principles in physics and philosophy to the psychology of sensation suggest? Hartshorne's answer, in a phrase, is the theory of the affective continuum. According to this theory, "the contents of sensation form an 'affective continuum' of aesthetically meaningful, socially expressive, organically adaptive and evolving experience functions" (PPS, 9). Thus the immediate data of consciousness comprise an ever-changing field of interpenetrating qualities, the qualities differentiated against a background of feelings socially involved with the feelings of others. The continuum is *affective*, since it is generated from contrasts of joy and sorrow, liking and disliking, etc., characteristic of feeling, and since the discrimination of contents within the continuum depends upon their comparison with recognizable forms of affectivity (PPS, 9–10). Affection, or feeling, forms a *continuum*, although holes do exist in any experienced "continuum" of sensory qualities; for the holes may be construed so as to reinstate the continuum of affection which, regardless of its seemingly hypothetical character, is empirically significant ". . . not only in its enlightening us with regard to what may exist even though as yet unobserved by us, but also in the fact that measurement of degrees of difference can take place as it were over the holes" (PPS, 10).

Now the theory of the affective continuum centers upon five conceptions (PPS, 6):

(1) The first conception is the application of mathematical continuity to sensation. As Hartshorne puts it: "The type of relation existing between colors, whereby one is connected with or shades into another through intermediaries, can be generalized so as to connect qualities from different senses (e.g., a color and a sound) or from different elementary classes (e.g., secondary and tertiary qualities)" (PPS, 6). Just as mathematical philosophy, in accord with Whitehead's method of extensive abstraction, considers points to be not elements of which the continuum is composed, but rather ideal limits of abstraction from the continuum, and just as physics, in line with Whitehead's criticism of the fallacy of simple location, views its ultimate physical elements of analysis as wave-packets into which the entire physical system focusses, so sensations are discriminable aspects of a continuum of felt quality. The point is illustrated by the color scale, which "is not an aggregate of colors, but a qualitative unity. . ." (PPS,

42). "Colors are thus aspects of the one fact which is 'color'"
(PPS, 43). What is true for color is true for all sensations, including those of different senses, and for all feelings as well.
Citing empirical evidence which Berkeley had noticed long
ago, Hartshorne observes: "Heat passes into practically pure
pain as orange into red... Thus, qualities from two senses,
one of them by all natural human conviction akin to the
'tertiary' or affective factor of sense experience, have ob-
servably the same continuity of nature as qualities from one
sense" (PPS, 50).

(2) The second conception is the primacy of aesthetic
meaning, or affective tone (Whitehead's "feeling-value").
"The 'affective' tonality, the aesthetic or tertiary quality,
usually supposed to be merely 'associated with' a given sensory quality is, in part at least, identical with that quality, one
with its nature or essence. Thus, the 'gaiety' of yellow (the
peculiar highly specific gaiety) is the yellowness of yellow"
(PPS, 7). Viewed as a subclass of feeling tones, sensory qualities are species for which feeling is the genus (PPS, 179).
Moreover, the development of sensations out of feeling is, for
Hartshorne, tantamount to asserting that "sensation is what
feeling becomes when externally localized in phenomenal
space," an assertion which he claims is supported by empirical evidence (PPS, 135).

(3) The third conception emphasizes the fundamental
social character of experience. "Experience," maintains
Hartshorne, "is social throughout, to its uttermost fragments
or 'elements.' Its every mode is a mode of sociability" (PPS,
8). Since, for Hartshorne, the sociality of experience entails
panpsychism, this conception of the social structure of experience will be considered again when panpsychism and
social realism are discussed. Here it suffices to state that the
sociality of experience signifies, in Whitehead's phrase, "the
feeling of feeling." (PPS, 193). Just as within the individual
self simple memory testifies to the feeling of past feeling, so the
self, according to Hartshorne, feels the feelings of the cells
that comprise its organism.[1] The extension of the principle

---

[1] Charles Hartshorne, "The Social Structure of Experience," *Philosophy*,
XXXVI (April and July 1961), 97–111.

of "the feeling of feeling" beyond human selves to God gives rise to the speculation that God in his self-enjoyment feels the feelings of all creatures. In this sense, "(t)he world may be conceived as the necessary specifications of the theme 'feeling of feeling'" (PPS, 208).

(4) The fourth conception has to do with biological adaptiveness. "The intrinsic natures of sensory qualities, and not merely the order and correlations in which they occur, express organic attitudes, or tend, of themselves, to incite modes of behavior; and these modes may be appropriate or useful, in relation to the physical circumstances generally accompanying the occurrence of the stimuli productive of the respective sensations" (PPS, 8). That sensations of pleasure and pain play a significant role in the adaptation of the organism to its environment is a commonplace, though the role of other sensations, e.g., colors, is less clear. Hartshorne, however, points out that these other sensations are also triggers to modes of behavior. Indeed, he argues that definite associations hold between particular modes of feeling and behavior and particular sensations of color or sound, and he concludes from this that psycho-physical associations are more fundamental in the life of the organism than associations of ideas (PPS, 250ff).

(5) The fifth conception pertains to evolution from a common origin. "The first appearance of a given quality at a certain stage in evolution is not a pure 'emergence' (though it has an emergent aspect) of the quality, unrelated to the previous state of nature, but is intelligible in much the same fashion as the appearance of a new organ" (PPS, 8). On this point – namely, the evolution of sensory qualities from a common origin, Hartshorne's indebtedness to Peirce's category of firstness is pronounced. Thus a continuum of sheer, undifferentiated, indeterminate, vague feeling becomes specified and determinate, through a process of objectification, into particular sensory qualities (PPS, 207–208).

II. PANPSYCHISM AND SOCIAL REALISM

The theory of the affective continuum, an elaboration of

the Whiteheadian metaphysics of events as feelings of feelings, is linked in Hartshorne's thinking with the explicit advocacy of panpsychism. Here the influences on Hartshorne's philosophy originate in Leibniz as well as in Whitehead, and to a lesser extent in Fechner. In its simplest form panpsychism asserts that "... all the pronounced units of matter are living" (BH, 166). The plausibility of panpsychism hinges, in large measure, upon the discarding of the concept of dead, static matter in physics, and, with promptings from biological science, upon the general adoption of the cell theory of reality. The discovery that the individual organism consists of living cells becomes the principle, which Leibniz employed first with considerable success, whereby material objects at the macroscopic level are construed to be colonies or aggregates of sentient microscopic monads. Within the organism, it is held, "... a cell is not only a living but a sentient organism" (PPS, 244). Against the prospective objection that the cell cannot feel because it has no nervous system, Hartshorne's retort is straightforward: The cell performs functions of digestion and reproduction without the appropriate organs. Similarly it may feel without the nervous system. "Lack of explicit organ does not spell lack of function, but primitive form of the function" (PPS, 244). The ascription of sentience to cells within the organism suggests, furthermore, the consideration of the simplest physical entities, electrons and other atomic particles, as centers of feeling (PPS, 249–250). That such entities are simple compared with higher animals does not preclude that they possess mind and will and feeling; "it merely means a low degree of complexity, and hence it is contrasted, not to mind and feeling and will in general, but to complex types of mind and feeling and will" (MVG, 214).

The Leibnizian-Whiteheadian principle of interpreting macroscopic objects and organisms and even human persons as societies of simpler, microscopic sentient beings, is the key to Hartshorne's advocacy of social realism within the framework of panpsychism. The mistake of Fechner, despite the high praise Hartshorne bestows upon him, springs from his inability to grasp the societal principle, an inability that leads him to treat composite organizations devoid of personal order as "compound individuals" and results in the bizarre and discredited panpsychism which attributes souls to plants and other macroscopic physical objects such as planets (PSG, 255a). The societal principle in Hartshorne's philosophy, moreover, signifies that feeling, identified with the very stuff of consciousness and universally applied of all being, is social. The feelings of an organism comprise complex feelings of the simpler feelings of the particular sentient beings, cells, that make up the organism; and these feelings in turn are feelings of the feelings of the yet simpler physical entities that constitute the external environment stimulating the cells of the organism. There is, then, a steady flow of feelings from the lower to the higher orders of organisms, even though some organizations of cells fall short of the high degree of unity necessary for personal order and integral consciousness. But, to return to the present point, the social structure of feeling is well illustrated by a consideration of color perceptions. The perception of a specific color, say red, involves the supposition that the nerve cells are qualified by this color. But what can a nerve cell so qualified be like? In answer to this sort of question Hartshorne, like Whitehead, breaks with the bifurcation of nature into objective physical primary qualities and subjective mind-dependent secondary qualities, and adhering to a social realism which imputes a continuity of felt quality between the organism and its environment, he writes:

Red is a mode of feeling value, describable in terms of the dimensions of social affectivity as such. The nerve cells have feelings also determinable on these dimensions. The relation between red as we see it and red as it is in the nerves is the relation between the individual units of a complex of feelings and the complex as a single over-all quality (PPS, 248).

Now in Hartshorne's philosophy, the case for panpsychism rests upon six arguments (RSP, 78ff):

(1) The first argument is the argument from *causality*. According to this argument, the relations between past events and present ones, relations exemplified by causality, require the persistence of the past into the present. This persistence

can be understood, Hartshorne claims, only if we suppose the pervasive presence of experiences akin to memory. The sole alternative to the supposition either of rudimentary memory experiences in seemingly physical entities or of an absolute cosmic memory which forgets nothing, in order to guarantee the causal order in nature, is, Hartshorne insists, the positivistic denial of causality (RSP, 78–79).

(2) The second argument is the argument from *unity within diversity*. According to this argument "apart from subjects" there is no principle for the one-in-the-many. Space and time and physical matter fail to provide the requisite unity (RSP, 78–79).

(3) The third argument is the argument from *contrast between particular and universal, actual and potential*. Although this contrast, or polarity, is a dominant characteristic of all being, it is intelligible, contends Hartshorne, only with respect to subjects with desires and consummations (RSP, 79–80). This argument is closely related to the fifth argument below.

(4) The fourth argument is from *the nature of quality*, and it involves concepts related to the affective continuum which I have already discussed. Briefly, this argument maintains that, since qualities qualify the object perceived and the nerve cells of the percipient organism, they indicate the presence of modes of feeling in both. "Thus all *known* qualities are actually qualities of feeling, whatever else they may be, and all knowable qualities are potentially qualities of feeling" (RSP, 80).

(5) The fifth argument is the argument from *the realization of possibilities*. Realization is a process of deciding between possibilities, of choosing some and rejecting the others that are incompatible with the ones selected. This is tantamount to creative choice, although prevalent throughout the natural cosmos, and creative choice, Hartshorne insists, ". . . seems totally unintelligible in the mere non-subject" (RSP, 82).

(6) The sixth argument is the argument from *intrinsic value*. According to this argument, intrinsic value inheres in beings with lives and feelings that may excite the interest of others and be enjoyed by them. Such beings must be subjects in the panpsychical sense. To be interested, even in what

common sense presumes to be dead matter, is, tacitly perhaps, to assume that values inhere in it, that, in other words, it is alive and feels. "The only intelligible conception of direct derivation of value from an object is," observes Hartshorne, "that the object has value to give, and this means, has its own values, its own life and feeling, and thus is some sort of subject" (RSP, 82).

In addition to six arguments to clarify and justify panpsychism, three important implications of the doctrine are discernible: in the philosophy of science, in the theory of knowledge, and in the ontology of individuality.

In the philosophy of science panpsychism means the unity of the sciences, but of an unusual kind. "(P)anpsychism is the doctrine that comparative psychology as psychophysiology. . . will ultimately include physics as the simplest branch" (BH, 179). The mind-body interaction and the subject-object relation, which since Descartes have given rise to baffling philosophical problems, are dealt with effectively by Hartshorne when he treats the relations between mind and body, subject and object, as consisting in bonds of organic sympathy, inasmuch as mind and body, subject and object, are societies of sentient entities sharing each others' feelings. Now this solution to what has been in effect *the* problem of modern philosophy is a form of idealism, since it requires subsumption of the allegedly physical under the psychical. This idealism is consistent with Hartshorne's speculative method. Being is regarded as a system of infinite variations, the values of which are concretely described by the psychological concepts of feeling, willing, thinking, remembering (BH, 115–118). As Hartshorne puts it: "We can generalize beyond human experience only by generalizing 'experience' itself beyond the human variety" (BH, 122). Of course, it should be made clear that Hartshorne does not intend to eliminate physics, but to retain it as a science confined to the simplest entities. "(P)hysics is only the behavioristic aspect of the lowest branch of comparative psychology," while, in view of the societal structure of reality, ". . . all psychology is in some sense social psychology, so that the final empirical science will be generalized comparative sociology" (MVG, 160).

In the theory of knowledge panpsychism means realism, but this realism is epistemological, and paradoxically it is compatible with and, indeed, culminates in metaphysical idealism. This theory, tantamount to a synthesis of idealism and realism, consists in the joint assertion of four theses:

(i) The principle of *Objective Independence:* "An 'object,' or that of which a particular subject is aware, in no degree depends upon that subject" (RSP, 70).

(ii) The principle of *Subjective Dependence:* "A 'subject,' or whatever is aware of anything, always depends upon the entities of which it is aware, its objects" (RSP, 70).

(iii) The principle of *Universal Objectivity:* "Any entity must be (or at least be destined to become) object for *some* subject or other" (RSP, 70).

(iv) The principle of *Universal Subjectivity:* "Any concrete entity is subject, or set of subjects. . ." (RSP, 70).

Following Whitehead, Hartshorne calls the theory which asserts these four theses "reformed subjectivism"; he also calls it "'societism' for it amounts to a social theory of reality" (RSP, 71).

In ontology panpsychism means the doctrine of "the compound individual." Whitehead's reformed subjectivist principle, adopted by Hartshorne, entails the consideration of macroscopic entities as consisting of microscopic subjects. Now the difficulties of formulating a defensible theory of substantial individuality, while acknowledging a plurality of individuals both internally and externally related some to others, are writ large over the history of philosophy. Whereas it is customary to locate the source of these difficulties in the alleged incompatibility between substance and process, an incompatibility which Hartshorne denies repeatedly, these difficulties really stem from the historic inability of philosophers to conceive how simpler substances are related, i.e. compounded, to form more complex substances without either losing the individuality of the components or attenuating the individuality of the compounds. Thus the history of philosophy divides, on the one hand, into a pluralism of discrete, unrelated individualities, atoms or monads, and, on the other hand, into a monism of absolute, undifferentiated substantial being. Neither alternative can be tolerated by immediate experience, which, as we have seen, is social in structure and involves the feeling of feeling, evincing the togetherness of microscopic individuals within more complex organisms. Although, of course, some organizations of simple individual entities, e.g., plants and stones, fall short of the high degree of personal order and consciousness requisite for true individuality and are consequently mere composites, there are, Hartshorne maintains, compound individuals. But the acknowledgement of compound individuals had to wait until Whitehead's philosophy. As Hartshorne writes: "In the 'cell theory' or 'philosophy of organism' of Whitehead we have nothing less than the first full-blooded, forthright interpretation of the cellular model (passing over the not much less adequate version found in Peirce's theory of the categories, and his doctrine of synechism, both of which conceptions have advantages not entirely paralleled in Whitehead's system). The theory of the enduring individual as a 'society' of occasions, interlocked with other such individuals into societies of societies, is the first complete emergence of the compound individual into technical terminology" (CI, 211).

III. PANENTHEISM

From the idea of the compound individual to the idea of God the transition is rapid and direct, for God is the maximal compound individual. The idea of God is the idea of a "'supreme' or 'highest' or 'best' individual (or superindividual) being. As a minimal definition, God is an entity somehow superior to other entities" (MVG, 6). At least this much theology demands, while religion adds that God must be a being worthy of worship. To be adequate, then, theology must expound and defend a concept of God which preserves the values which religion emphasizes; it must ". . . express and enhance reverence or worship on a high ethical and cultural level" (DR, 1). According to Hartshorne, only panentheism affords a theological concept of God which possesses the desired religious value. Although Hartshorne credits many ancient and modern philosophers, including Plato, Sri Jiva, Schelling, Fechner, Peirce, Iqbal, Berdyaev, and Weiss, with shaping theological panentheism, the high esteem he bestows upon Whitehead is emphatic. In the comment on White- head's theory in *Philosophers Speak of God*, we find: "It is impossible to avoid a feeling of impertinence in attempting to comment on thinking so great as this. Not in many centu- ries, perhaps, has such a contribution been made to philo- sophical theism" (PSG, 282b).

Hartshorne defines panentheism by means of a consider- ation of the primary attributes predicated of God, dis- tinguishing it from its main rivals, of which the foremost are traditional theism and pantheism. Panentheism affirmatively answers five questions pertaining to the concept of God:

"Is God eternal? Is he temporal? Is he conscious? Does he know the world? Does he include the world?" (PSG, 16b) Hence Hartshorne's panentheism asserts: (1) that God is "Eternal – in some. . . aspects of his reality devoid of change, whether as birth, death, increase, or decrease. . .," (2) that God is "Temporal – in some. . . aspects capable of change, at least in the form of increase of some kind. . .," (3) that God is "Conscious, self-aware. . .," (4) that God is "Knowing the world or universe, omniscient. . .," and (5) that God is "World- inclusive, having all things as constituents. . ." (PSG, 16b). God, therefore, is "The Supreme as Eternal-Temporal Consciousness, Knowing and including the World" (PSG, 17a). And this conception of deity not only guarantees the values upon which religion insists, but also conforms to the most rigorous logical analysis and finds empirical confirmation in the theories of nature supported by contemporary science.

Rival conceptions of deity are differentiated from pan- entheism by their omission of one or more of the five attributes ascribed to God. Thus the history of philosophy witnesses eight alternative theological doctrines other than Hartshorne's panentheism: (1) God as an eternal consciousness, neither knowing nor including the world, as in Aristotelian theism; (2) God as an eternal consciousness that knows but excludes the world, as in the classical theism of Philo, Augustine, Anselm, Aquinas and Leibniz; (3) God as "the Eternal be-

yond consciousness and knowledge," as in Plotinus' ema-
nationism; (4) God as "Eternal Consciousness, Knowing and
including the World," as in the classical pantheism of Spinoza
and Royce; (5) God as "Eternal-Temporal Consciousness,
Knowing but not including the world," as in the temporalistic
theism of Socinus and Lequier; (6) God as "Eternal-Temporal
Consciousness, partly exclusive of the World," as in the
limited panentheism of James and Brightman; (7) God as
"wholly Temporal or Emerging consciousness," as in the
cosmology of S. Alexander; and (8) God as "temporal and
non-conscious," as in the theology of Wieman (PSG, 17).

Hartshorne's conception of God, unlike its alternatives,
fastens on the juxtaposition of predicates representing
eternity and omniscience and of those representing tempo-
rality and world-inclusiveness. It is unquestionably the most
comprehensive conception of God, but its comprehensiveness
heightens the contrast within God between an aspect which
is absolute and unchanging and an aspect which is relative
and changing. This contrast, akin to Whitehead's distinction
between the primordial and consequent natures of God,
entails, as Hartshorne contends, "Divine Relativity." In this
sense, panentheism is Surrelativism, with God the Compound
Individual conceived as the Process. According to Surrelativism, ". . . the 'relative' or changeable, that which depends
upon and varies with varying relationships, includes within
itself and in value exceeds the non-relative, immutable, independent, or 'absolute,' *as the concrete includes and exceeds the
abstract*" (DR, vii). From this ". . . it follows that God, as
supremely excellent and concrete, must be conceived not as
wholly absolute or immutable, but rather as supremely-rela-
tive, 'surrelative,' although, or because of this superior relativity, containing an abstract character or essence in respect
to which, but only in respect to which, he is indeed strictly
absolute and immutable. . ." (DR, vii).

Now for Hartshorne none of the eight conceptions of God
depicted above quite equals his own panentheistic conception
of God as the supreme eternal-temporal consciousness,
knowing and including the world. None attains tne fullness
of being crystallized in the dipolar theory of the divine rela-

tivity. It is no surprise, then, that the history of theology is divided between classical theism and pantheism, both regarding God in monopolar terms. Whereas classical theism stresses the absoluteness and transcendence of God, radically separates God from the World, and guarantees his immutability by excluding all passivity and becoming from him; pantheism identifies God with the total system of all changing things, and consequently denies any absolute, transcendent, or independent side to his nature. Thus as classical theism favors a conception of God which fails to portray his supreme perfection adequately, since the cosmos is greater than God in that it contains God and the World; pantheism never succeeds in grasping that aspect of God's nature which is absolute in its perfection and independent of the vicissitudes of becoming. The pantheist and the traditional theist seize upon one set of contrary attributes and disregard the other set. As Hartshorne says: ". . . common to theism and pantheism is the doctrine of the invidious nature of categorical contrasts" (PSG, 2a). Pantheism and classical theism have, in effect, generated an artificial dilemma for theology, because both ". . . have assumed that the highest form of reality is to be indicated by separating or purifying one pole of the ultimate contrasts from the other pole" (PSG, 2a).

The way around the dilemma, Hartshorne argues, is to abandon the principle of monopolarity upon which it is based, and to substitute the principle of polarity.[1] Borrowed from Morris Cohen,[2] the principle of polarity dictates that ". . .ultimate contraries are correlatives, mutually interdependent. . ." (PSG, 2b). The adoption of this principle in theology signifies that ". . . all contrasts. . . fall within God

[1] Professor Robert C. Whittemore informs me that panentheism need not be restricted, as Hartshorne demands, to a dipolar conception of God. And so, Whittemore maintains, Hartshorne's classification of alternative conceptions of deity in his logic of panentheism collapses. Spinoza, for example, may be considered a panentheist rather than a pantheist. For the detailed argument on this and related points, see Whittemore's forthcoming book, *In God We Live; A Critical History of Panentheism.*

[2] Morris Cohen derived the principle of polarity from Wilmon H. Sheldon in the latter's seminars at Columbia University in the late 1890's. For a discussion of the role of polarity in Sheldon's thought, see my essay, "Wilmon H. Sheldon's Philosophy of Philosophy," *Tulane Studies in Philosophy*, VII (1958), 111–128.

(since, in one aspect of his reality, he is the most complex and inclusive of all beings), but each contrast is in God in its own appropriate way. . ." (PSG, 15a). For Hartshorne dipolarity points the way beyond the monopolarity of both pantheism and traditional theism to the doctrine of panentheism. In panentheism God is both immanent in and containing the world of changing, individual, dependent beings, and an eternal, absolute, independent being transcending the world (DR, 90).

Furthermore, God is conceived at last to be both personal and social, a being who changes yet has absolute eternal aspects. Indeed, the personality and sociality of God are mutually implicative, since to be a person is to be "qualified and conditioned by social relations, relations to other persons. . ." (DR, 25). Traditional attributes, such as omniscience and love, make sense for God only if he is a social being related to others. The traditional presumption that God knows and loves other beings but is unrelated to them, although they, in their dependent status, are related to him, paradoxically reverses, as Hartshorne insists, the ordinary view that the objects of knowledge and of love make a difference to the knower and lover, perhaps more so than the knower and lover make to them. It is common sense that the denial that God is related to his creatures is equivalent to the denial that he knows or loves them. To know and to love are to be related, and the relation involved in this case is the relation of inclusion (DR, 16–18). When God knows or loves his creatures, he includes them eminently, not in his essence, of course, but within his total being.

At this point a serious problem arises: If God includes beings that are imperfect, then he is imperfect and so not worthy of worship. On the other hand, if he excludes any beings, then the totality which includes them and him is greater than he, and he is no longer supreme. The solution Hartshorne offers is ingenious. The perfect being is defined as a being which is unsurpassable except by itself (MVG, 8ff and 342ff). The perfect being, God, is a being for whom "self-superiority is not impossible," so that ". . . the perfect is the self-surpassing surpasser of all'" (DR, 20). God embraces all the value there is; he lacks none. This does not mean, however, that more value will not come to be in the future; but if and when it does come to be, God will embrace this new value, too. God, therefore, changes, not of course in those abstract aspects which are eternal, but so far as he is related to and includes changing things.

Here, again, the attribute of omniscience illuminates the dipolar nature of God. Basically, the omniscience of God is eternal, and always God knows all there is to know – namely, whatever was and whatever is. But besides this absolute aspect of God as eternally omniscient, there is, according to Hartshorne's analysis, a relative, changing aspect. Though God is eternally omniscient, his knowledge changes. As new events occur each moment, he knows them; but prior to their occurrence, when they are not yet actual, God's knowledge does not grasp them as actual, since to do so would be to falsify what they are; God does know them as possibilities, but as possibilities they subsist within an indefinite range of indeterminate alternatives. No possibility is definitely actual before it is realized, and nothing destines one set of possibilities rather than another for realization prior to the actual, concrete course of becoming (MVG, 99ff). Hence God, though eternally and absolutely omniscient, does not know what will happen tomorrow until it *does* happen, and his omniscience has both an eternal, absolute and a temporal, relative aspect. Since what is found true for omniscience is true also for the other attributes of God, it follows that God, a perfect being, yet is a changing being.

To the problem of demonstrating the existence of God, Hartshorne has devoted considerable thought. It should be obvious that the arguments for panpsychism, which I have already discussed, are germane, insofar as they point to the existence of a universal subject. And the pages of Hartshorne's voluminous writings are dotted with passages which reflect the perspective afforded by panpsychism on demonstrating the existence of God. The following passage is illustrative:

How do we know that God exists? The universe must have some primordial and everlasting character, as the ultimate subject of change. The past being immortal, there must be a complete cosmic memory, since the past in the present *is* memory. The future being predictable, there must be a world anticipation; for the future as fact in the present is anticipation. Also, action implies the faith that at no time in the future will it ever be true that it *will* have made no difference whether the action was well-motivated or ill. This condition is met by the affirmation of a God who will never cease to treasure the memory of the action and of its results (CI, 218).

Later in the career of his authorship Hartshorne has remarked that all the arguments for God's existence "amount to this: that the proposition, 'There is a supremely excellent being, worthy of worship,' expresses fundamental or categorical aspects of experience and thought, while the denial of this proposition contradicts such aspects. There can be as many arguments for God as one can distinguish fundamental aspects of experience and thought..." (PSG, 24b). Hartshorne suggests a list of six arguments: "the aesthetic argument, the ethical argument, the epistemological ('idealistic') argument, the design argument, the cosmological argument, and the ontological argument" (PSG, 25a). Although these arguments have been operative since the beginning of theology, and although they have been corrupted by their linkage with the artificial dilemma between traditional theism and pantheism, they can, Hartshorne believes, "... be given a more exact and perspicuous form than has hitherto been given them. But this is a subject for another occasion" (PSG, 25b).

The panentheistic conception of God, argues Hartshorne, is not only logically exact and theologically full; it is also the only conception of God worthy to inspire man's worship. At last man's vision of God in religion is fulfilled in his vision of God in theology. The "Divine Relativity" introduces a God who is active and passive, who creates and enjoys his creatures but suffers and sorrows with them as well. "Here," Hartshorne comments with a nod to Berdyaev, "the Christian idea of a suffering deity – symbolized by the Cross, together with the doctrine of the Incarnation – achieves technical metaphysical expression" (PSG, 15b). And so God is both personal and social. Hartshorne relates how once in private conversation Whitehead had described God "as a 'society of occasions' (with personal order)" (DR, 30–31), and from this description of God to the panentheistic conception of God as the most perfect compound individual the distance is brief. Or, to employ different words, the conception of God Hartshorne offers, in accord with "Whitehead's supreme conception. . . of a society of actual occasions, related one to another by the sympathetic bond of 'feeling of feeling'" (DR, 29), has as its reverse side the panpsychical conception of the world as the divine organism. The doctrines of the affective continuum, of panpsychism and social realism, and of panentheism are expressions in technical philosophy which translate St. Paul's dictum: "We are members one of another." (BH, 123).

THE METAPHYSICS OF WHITEHEAD'S FEELINGS[1]

ROBERT C. WHITTEMORE

On page 246 of A. N. Whitehead's *Process and Reality* occurs this statement:

The primitive form of physical experience is emotional – blind emotion – received as felt elsewhere in another occasion and conformally appropriated as a subjective passion. In the language appropriate to the higher stages of experience, the primitive element is *sympathy*, that is, feeling the feeling *in* another and feeling conformally *with* another.

"The feelings," he tells us later on, "are inseparable from the end at which they aim; and this end is the feeler. The feelings aim at the feeler, as their final cause. The feelings are what they are in order that their subject may be what it is. Then transcendently, since the subject is what it is in virtue of its feelings, it is only by means of its feelings that the subject objectively conditions the creativity transcendent beyond itself" (PR 339). If Whitehead is right, then the first chapter of any metaphysics of the modern world must be a theory of feelings. The foundations of cosmology are laid in psycho-physiology, *if* Whitehead is right.

That he is, in fact, right, the current literature of psychology and philosophy bears witness. During the past twenty years, no one, psychologist or philosopher, has, so far as I have been able to determine, seriously challenged his assumption that "the basis of experience is emotional." Such criticism as there has been, and frankly there has not been very much,[2] consists of suggestions for minor revisions of the

[1] A paper presented to the Symposium on Whitehead's Psychology at the 53rd annual convention of The Southern Society for Philosophy and Psychology, Atlanta, March 31st, 1961.

[2] See John K. McCreary, "A. N. Whitehead's Theory of Feeling," *Journal of General Psychology*, Vol. 41 (1949), 67–68; also A. H. Johnson, "The Psychology of Alfred North Whitehead," *Journal of General Psychology*, Vol. 32 (1945), 175–

feeling-theory. The rest is silence. By default, Whitehead's "philosophy of organism" [1] has become the new psychophysiological orthodoxy.

Now this is not, I submit, a happy situation. A perennial distrust of orthodoxy has ever been in both our disciplines the mark of progress. To fan this distrust is an obligation shared by all. The problem is, how, given such default of basic criticism, this may best be accomplished. The answer, I suggest, lies in attending rather more carefully to the metaphysical consequences of the theory. For if our previous assertion is correct, if Whitehead's cosmology is implicit in his psychology, then adoption of that psychology must imply adoption of that cosmology which is its consequence. And yet, how few have recognized this; let alone accepted it. To judge by their evidences, few indeed. Certainly this is inconsistent. If you accept Whitehead's psychology, you surely have no sound reason to deny the cosmology and theology it implies. In brief, I offer this choice: take issue with this theory of feelings or stand ready to reverence strange gods.

II

To those inclined to think in terms of subject and predicate, the former alternative must undoubtedly commend itself; the more so in light of the difficulties inherent in Whitehead's conception of the subject (feeler). For not only does his theory of feelings require us to abandon all subject-predicate forms of expression as inadequate to the description of the Real; it lays upon us the burden of understanding just how it is that an actual entity can be a final cause. The ultimate subjects, the final facts, are, he tells us, all alike, acts of experience, events in process of enactment, feelings aiming at their feeler.

212; and Victor Lowe, "William James and Whitehead's Doctrine of Prehensions," *Journal of Philosophy*, Vol. 38 (1941), 113-126. Full scale studies of the salient features of Whitehead's theory of feelings are offered by Charles Hartshorne, *The Philosophy and Psychology of Sensation* (Chicago: University of Chicago Press, 1934), and by William A. Christian, *An Interpretation of Whitehead's Metaphysics* (New Haven: Yale University Press, 1959).

[1] "The philosophy of organism is a cell-theory of actuality. Each ultimate unit of fact is a cell-complex, not analysable into components with equivalent completeness of actuality" (PR 334).

Subjects (feelers), then, are processes with purposes, and not, as in ordinary usage, entities enjoying careers. Internally considered, they are final rather than efficient causes, and the mere fact that Whitehead chooses to term them actual entities makes them no less final or more efficient. In the language of *Process and Reality:*

> The determinate unity of an actual entity is bound together by the final causation towards an ideal progressively defined by its progressive relation to the determinations and indeterminations of the datum. The ideal, itself felt, defines what 'self' shall arise from the datum; and the ideal is also an element in the self which thus arises (PR 228).

To say that the ideal is an element in the self rather than the self itself in no way lessens our problem; for in the absence of any other element available to serve as subject (feeler), we are still faced with conceiving an ideal entitatively. If an entity is, in truth, a purposive process; what is it in itself antecedent to its achievement of its end?

That it is, in fact, *something* in itself, Whitehead plainly holds: "Every actual entity, in virtue of its novelty, transcends its universe, God included" (PR 143). The possibility that the entity might be its own efficient cause, he just as clearly rules out: ". . . efficient causation expresses the transition from actual entity to actual entity; and final causation expresses the internal process whereby the actual entity becomes itself" (PR 228). Has he forgotten Hume's dictum that all causes are of the same kind? [1] At all events, he seems not to have considered the line of solution it suggests. It remains then that an actual entity, genetically regarded, is only and always a final cause, and this being so, it follows that the ultimate constituents of the universe are all alike – *final* causes. Is such a universe conceivable? Can there be any such thing as a real plurality, not to say infinity, of *final* causes? If we allow some doctrine of pre-established harmony,

---

[1] ". . . all causes are of the same kind. . . in particular there is no foundation for that distinction, which we sometimes make betwixt efficient causes, and causes *sine qua non*; or betwixt efficient causes, and formal, and material, and exemplary, and final causes." David Hume, *A Treatise of Human Nature* (Oxford: Clarendon Press, 1949), Section XIV, p. 171.

defined as the primordial operation of deity on creativity, perhaps. Otherwise, I think not. We may, if we please, reject Whitehead's theory of the feeler as final cause; but if we accept it, we must, I submit, accept also that God which alone makes the theory intelligible and workable.

A similar solution suggests itself with regard to the difficulties involved in Whitehead's conception of the transmission and transmutation of feelings. For instance, that there is feeling of feeling, Whitehead firmly believes. On the other hand, if Christian's analysis [1] is sound, and I think that in the main it is, feeling of feeling "does not involve any common immediacy, because the feeling felt is objectified. The feeling felt exists, for the feeler, not in the mode of subjective immediacy but in the mode of objective immortality. There is no sharing of immediacy." [2] But if there is no literal transference of immediate feeling from the past to the present event, how shall we account for the continuity of process? As above, the answer seems to be a direct appeal to the efficacy of God, as, incidently, does also any explanation of that undescribed [3] transmutation essential to the emergence of every higher level of experience. In sum, there is no phase of feeling-theory immediately intelligible in itself. Only God can make the theory go. [4]

III

For those unwilling to admit that psychology finds its meaning and completion in theology, there are two courses open: they can seek a secularist theory of feelings, or, failing

[1] William A. Christian, *An Interpretation of Whitehead's Metaphysics*, (New Haven: Yale University Press, 1959).

[2] Ibid., 154–155.

[3] Nowhere in his later works does Whitehead offer the reader any detailed explanation of the genetics of transmutation. Like creativity, it is simply presupposed. Are we, then, to think of transmutation as something absolutely ultimate, intrinsically indescribable? Probably so.

[4] According to Christian, Whitehead's "theory of actual occasions logically requires his doctrines of God. The subjective aim of an occasion, on which its unity, its individuality, and its social transcendence all depend, requires an explanation. And the only explanation available on Whitehead's principles is one requiring his doctrine of God. His doctrine of God is in this way required by the theory of actual occasions, and is not a merely decorative addition to that theory." *Interpretation*, 310–311.

to find any such, they can abandon the assumption that the basis of experience is emotional. As for the first course, the outlook is not promising. If there is any such thing as a secularist theory of feelings, i.e., a theory whose conception of causation is of one purely efficient and in no wise final, I must confess my ignorance of it. Every cosmology which takes evolution seriously, must and does presuppose some quality of purposiveness, for evolution minus purpose would not be evolution but simply change. Moreover, it is at least open to question that purposiveness of any sort is compatible with secularism of any psychological type. Certainly, in philosophy the admission of purposiveness has always been and is today the standard preface to recognition of deity in some form.

You will ask: why must we admit purposiveness at all? Deny this, and deity dissolves like the mirage it probably is. Just so; but can you be sure that such a denial is not simply another case of all coherence gone? In other words, can anyone who wishes his psychology to be taken seriously refuse the cosmo-theological question? If Whitehead is right, and I for one believe he is, "all constructive thought on the various special topics of scientific interest is dominated by some such (cosmo-theological) scheme, unacknowledged, but no less influential in guiding the imagination" (PR x).

It remains to remark the one course still unconsidered, i.e., the abandonment of the original assumption of the emotional basis of all experience. Suppose this done? What then? Shall we return to sensationalism, ignoring as we do so, the existence of all that is not anthropomorphically clear and distinct? Shall we wait upon behaviourism to account for all those difficulties that sensationalism leaves untouched? With all due respect to the partisans of these heroic electives, I cannot bring myself to believe that either promises much as an alternative to Whitehead's theory of feelings. That the theory has its loose ends, I freely concede; nonetheless, for all of its shortcomings, it remains, as noted above, that of all the options before us, it alone is true to the spirit of modern science. The fact cannot be shrugged: anyone determined to do justice to the doctrines of relativity and evolution ultimately has no choice but to begin here.